General Principles of Good Sampling Practice

General Principles of Good Sampling Practice

Neil T. Crosby and Indu Patel
Laboratory of the Government Chemist, Teddington

THE ROYAL
SOCIETY OF
CHEMISTRY
Information
Services

LGC
Laboratory
of the Government Chemist

VAM
VALID ANALYTICAL MEASUREMENT

A catalogue record for this book is available from the Brithish Library.

ISBN 0-85404-412-4

© Crown Copyright 1995

Published for the Laboratory of the Government Chemist
by The Royal Society of Chemistry,
Thomas Graham House, The Science Park, Cambridge CB4 4WF

Printed and bound by
Redwood Books Ltd., Trowbridge, Wiltshire

Preface

This document has been prepared under the Department of Trade and Industry Valid Analytical Measurement (VAM) Initiative which forms part of the UK National Measurement System. The 1991–1994 programme of the VAM initiative included a project which had the objective to bring together generic aspects of problems encountered in obtaining reliable and representative samples of bulk materials for analysis.

This report has been prepared by two members of staff of the Laboratory of the Government Chemist following an extensive literature survey covering a wide range of materials. Hence, the report concentrates on the principle of good sampling practice and does not give detailed instructions for the sampling of any one particular commodity.

The authors gratefully acknowledge the assistance received from other colleagues at LGC.

Contents

Acknowledgements

A preliminary draft of this report was circulated to a large number of individuals in a wide range of industrial concerns, academia, government, Public Analysts, Private Consultants, and miscellaneous organisations. Many of these individuals responded with helpful criticisms and suggestions for improvement. Where possible these comments have been incorporated into the final document.

The authors are most grateful to all those individuals who took the time and trouble to read the document. We believe that the final version is much improved as a result of the assistance we have received.

We are also grateful to Brian Currell (Greenwich University Press) for allowing us to use some extracts from their earlier publication on Samples and Standards (Analytical Chemistry by Open Learning).

Abbreviations

AMC	Analytical Methods Committee of the Royal Society of Chemistry, UK.
AOAC	Association of the Official Analytical Chemists International, USA.
ASTM	American Society of Testing Materials, USA.
BS	British Standards Institution, UK.
COSHH	Control of Substances Hazardous to Health, UK.
DTI	Department of Trade and Industry, UK.
VAM	Valid Analytical Measurement Programme, DTI, UK.

General Principles of Good Sampling Practice

1 Introduction

The Department of Trade and Industry's (DTI) initiative on Valid Analytical Measurement (VAM) is designed to improve the quality of analytical measurements and so facilitate the mutual recognition of data throughout the UK and Europe. Sampling forms an essential part of the VAM programme, since however carefully analyses are carried out, the results will be of limited value unless the portion of the sample taken for analysis is truly representative of the bulk material. Frequently, consignments as large as several hundreds of tonnes have to be examined whilst the final stage of the analysis can involve the injection of only a few microlitres of solution into an instrument; sampling is an important stage in bridging this enormous gap.

Sampling of materials is also required as many chemical methods of test are destructive. Hence examination of the entire commodity even if it were practicable would consume it all leaving nothing for commercial use. Indeed some chemical tests can only be done on small quantities of test material. Processes such as digestion or solvent extraction require increased times and volumes of reagents as the weight of test sample is increased, thus increasing the cost of analysis and the magnitude of the blank values which is particularly important in trace analysis. Whereas this creates few problems for homogeneous products, blends of materials varying widely in chemical composition or physical properties are often difficult to sample representatively. Hence, analyses carried out on small (gram) portions withdrawn from the bulk are likely to vary greatly from determination to determination. A single measurement would provide an erroneous value which could lead to incorrect (and costly) decisions being made.

One other important aspect of sampling is where measurements are made on a batch during production as part of quality control (process control monitoring). If possible this is achieved by on-line analysis using continuous monitoring equipment or, alternatively, by repeated sampling and separate measurement. These techniques are outside the scope of this volume but are discussed fully elsewhere. [1-3]

Over the years much attention has been paid to the analytical determination. Methods have been developed and validated by a large number of official bodies including AMC, AOAC, UK, European and International Standards Organisations as well as specialist Trade bodies. In recent years however more attention has been paid to 'Total Quality Measurement', including external accreditation and proficiency testing schemes for the independent assessment of laboratories. Certified reference materials are being developed continuously to ensure traceability of measurements in appropriate cases. However, much less attention has been paid to sampling, the critical first step in any examination of a consignment or product, although many trade standards do incorporate recommendations for sampling. There are also many publications in learned journals but the existing knowledge and expertise in sampling is widely scattered and often not readily accessible, especially to newly trained students leaving colleges of higher education. It is not possible to provide detailed instructions for the sampling of every possible type of product encountered in the commercial world in a publication of manageable size. Even if it were feasible, such a publication would be of interest to many people but few would be interested in the whole. Sampling is frequently entrusted to the lowest grade of employee, often with the minimum or no training in the principles of good sampling practice. Sampling and analysis are inextricably linked. Therefore it is desirable to provide the sampling personnel with adequate levels of training and supervision when undertaking any sampling tasks. There is little point in striving to improve analytical errors if they are better than about 45% of the sampling error. The overriding criterion is 'fitness for purpose'. Both sampling and analysis must be carried out in such a way that the final data obtained enable correct and sensible decisions to be taken.

Hence, the aims of the present project are:

- to develop generic principles of good sampling practice using existing knowledge,
- to validate these principles using case studies to demonstrate that the principles are comprehensive, and
- to create media to disseminate these principles, and set up accreditation schemes.

1.1 Aims of This Book

Since commodities to be sampled vary so widely in their chemical composition and physical properties, it is self evident that no one single procedure can be recommended to obtain a truly representative portion from a bulk material. Nevertheless, there are a number of general principles that can be applied to all types of sampling activity. These common features have been drawn together to produce a set of practical yet general guidelines. It is hoped that these guidelines will be used to inform the design of future sampling plans. Although there are aspects of sampling that are common to all situations, equally there are differences that can be important in individual circumstances. However, the reader will be able to check if his or her particular scheme satisfies the general guidelines. Individuals are best able to decide when or where divergence from the general scheme is justified based on their own experience and knowledge of the properties and characteristics of the product to be sampled!

This book attempts to summarise these general principles and demonstrate how they can be applied in a few illustrative case studies covering solids, liquids, and gases. Information is also presented on the types of equipment available for sampling and some key references to the literature. It is hoped that this basic knowledge will enable the reader to develop and improve his own sampling practice and specific requirements. The book summarises a number of key parameters that may need to be considered when drawing up a sampling plan and putting the plan into practice. Some of these parameters may not be important for the particular job in hand. Others may need to be studied and developed in much greater depth than has been possible in this publication. A series of check lists of aspects of sampling applicable to certain common sampling problems has been designed for use in the field. The book also offers some comments on the application of statistics to sampling methods (Appendix 1) and a glossary of the terms often found in the literature on sampling (Appendix 2). Safety is an important consideration for sampling officers and key aspects will be discussed, although precise recommendations can only be made by those conversant with the hazards of the site and the chemical nature of the product to be sampled. Other activities relevant to sampling include sub-sampling either on site or in the laboratory. These aspects will be considered in as far as they affect the procurement of a representative test sample for analysis.

1.2 Sampling Procedures

The act of sampling consists of two parts:

(i) The design of a sampling plan
(ii) The implementation of the agreed plan in a practical situation.

These two activities are quite separate and so will be treated individually in this publication. It is also likely that different personnel will be involved in the separate activities, although it is essential that good liaison exists between the two teams and also with the analytical scientists who will subsequently perform the tests in the laboratory. The sampling plan must consider not only the method of taking the sample, when and from where, but also any subsequent treatment prior to analysis in the laboratory.

The principal steps in sampling and analysis operations are shown in summary form in Figure 1. The first three of these steps will be considered in more detail in the following chapters.

Environmental sampling is largely outside the scope of this project but many of the principles of sampling bulk goods can be applied also to environmental sampling. The design of the sampling plan is of crucial importance in environmental sampling and a few factors for consideration will be highlighted.

Food surveillance for chemicals (additives, nutrients, and contaminants) present in the nation's food supply forms an important part of the Government's public health programme. Sampling is a critical part of food surveillance and so some general considerations will be made.

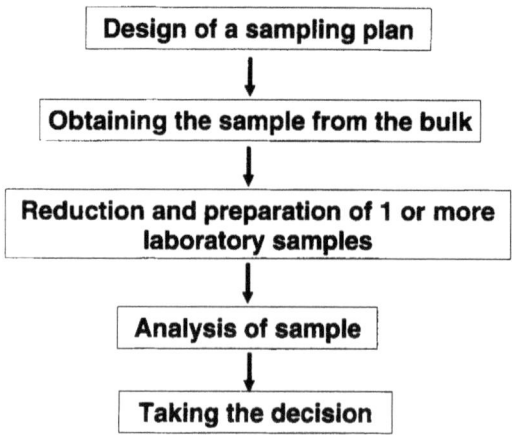

Figure 1

1.3 Other Publications

This book is one of a number of publications which are being produced as part of the VAM Sampling Strategies Project. These publications are complementary and form a range of products designed to address the subject of sampling from the best practice angle. They aim to highlight the fundamental importance of valid sampling on the analytical measurement process, and to provide those involved with sampling the information necessary to formulate valid, relevant sampling plans and protocols.

2 Sampling Plans

The design of a sampling programme must ensure that the information required by the objectives of the measurements are met. There is little or no point in routine implementation of a programme known to be incapable of providing this information. Thus, the objectives are the prime factor in governing the resources and effort required for sampling and analysis. Hence it is essential that the initial set of objectives is carefully defined.

Figure 2 summarises the sequential order of the factors to be considered when designing a sampling programme.

2.1 Objectives

Commodities to be sampled may vary from a consignment of several thousand tonnes (*i.e.* a whole boat load) to a single tin or packet of a foodstuff. In the case of a boat load, it is unlikely that the entire product will have been produced in a single batch. Hence, there may be batch to batch variations, in addition to the inherent inhomogeneity of the product itself. The objectives of sampling may

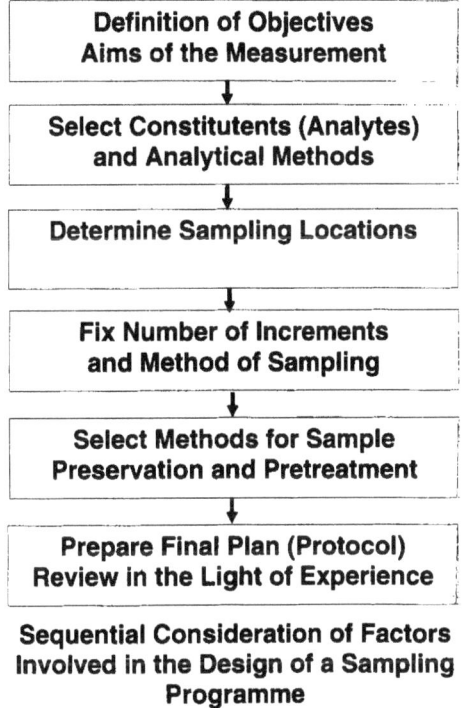

Sequential Consideration of Factors Involved in the Design of a Sampling Programme

Figure 2

therefore vary from a need to determine the homogeneity (or heterogeneity) of a product to the need to smooth out such variations, to determine the *average* composition of the bulk material or consignment.

The objectives must be written down in as clear and simple a format as possible and agreed by *all* parties to the work programme. The key questions to ask to test the validity of the objectives are:

(i) what use will the results obtained on the samples following analysis be put to, and
(ii) what decisions will be taken when the sampling and analytical work has been completed?

Some examples of possible decisions are given below:

1. Acceptance/rejection of a product for use in a manufacturing operation.
2. Assessment of the value of a consignment of goods followed by purchase.
3. Quality control of a final product before sale.
4. Evaluation of a product offered for sale against statutory or other quality criteria with a view to prosecution if found to be deficient.

Obviously, incorrect results in the above cases resulting either through mistakes in analysis or from incorrect sampling could cost an industrial company a large sum of money. In the case of a final product, this might also lead to prosecution, fines and unwelcome and costly publicity.

Enforcement and consumer protection agencies also need to bear in mind the principles of correct sampling, although statutory procedures may be prescribed in such cases. Environmental sampling is also critical, as decisions affecting the *health* of the population may depend on the veracity of the results obtained. Again, it is vital to ask the questions 'Will the aims of the measurements be achieved?' and 'What action will be taken following the results of a given monitoring programme?'.

Central to the definition of clear and concise objectives is an understanding of the uncertainty attached to the final analytical result. This uncertainty will arise both from the analysis and from sampling. The sampling plan must take this into account and be so designed to reduce the error to an acceptable level, bearing in mind the decision to be taken at the end of the study.

Equally, there is no point in designing an over-elaborate sampling programme when extreme accuracy is not required for the subsequent decision. For example, the sampling and analysis may be undertaken merely to ensure that a particular constituent is present within a given range. Alternatively, it may be necessary to establish that a given contaminant is absent, or present at a concentration no greater than say 100 mg kg^{-1}. Hence, in all cases the sampling plan must be assessed against the criterion 'fitness for purpose'. Changes may have to be made to the plan subsequently following experience and feedback from initial results.

Large amounts of time and other resources are consumed by sampling and analysis. It is clearly desirable that such resources are used with maximum efficiency. Time spent on designing the sampling plan will be well rewarded, particularly when the programmes are of a continuing nature. If the information required on quality is not carefully defined, it is clear that the measurement programme may be inappropriate or inefficient or both. Care is required to ensure that the cost of the measurement programme does not exceed the benefits of the results obtained. Statistical considerations (Appendix 1) may help to define the sampling plan in more quantitative terms, depending on the degree of accuracy required.

In summary the sampling plan when completed must provide answers to the following questions:

- What do we want to know?
- Why do we need this information?
- What happens to the results?
- What actions may follow?

Before these questions can be answered, it will be necessary to gather pertinent information so that the factors in Figure 1 can be resolved and agreed.

2.2 Analytes and Methods

Any sampling plan must define the analytes (or constituents) to be determined in the sample. The analytical methods to be used must also be agreed. Preference should be given to those methods which have been collaboratively tested and for which data on accuracy and precision are available. It is then necessary to check whether these data are sufficient for the job in hand. As a general case, uncertainty in analysis constitutes about one third of the total uncertainty incurred by sampling and analysis.[4] Hence, for inhomogeneous products where the sampling error is large, there is little value in trying to improve the analytical method. Better results will be obtained by the analysis of a larger number of samples. Where collaboratively tested methods are unavailable and in-house methods have to be used, performance criteria for the methods should be obtained, albeit on a more limited scale and assessed for suitability for the job in hand. The limit of detection will be of prime importance for trace analysis and may need to be specified in the sampling plan. In screening or surveillance work, the number of false positives (or false negatives) may have to be specified or controlled. The object of a screening programme is to test a large number of samples covering as wide a range as possible using rapid and cheap methods. Those samples giving a positive response can then be examined by more specific confirmatory or reference methods. Obviously, the number of such samples subject to re-test should be kept to a minimum. Equally, insensitive methods should not be used so that samples containing the analyte at a low level are recorded as negative (*i.e.* false negatives).

2.3 Sampling Locations and Increments

It is not always easy to identify the lot or batch of material to be sampled. A warehouse may be full of packages similar in shape, size, and appearance, but containing different materials. Heaps of materials, although piled separately, may be from the same or different batches. Hence, it is first necessary to define the lot or batch to be sampled. Secondly, it is necessary to define the number of incremental samples to be taken from the lot and how this is to be undertaken. If the batch consists of a number of packages or cans, the number to be sampled will need to be stated. It is also necessary to describe how this number is to be drawn from the total number in such a way that each package or can has an equal chance of being selected for the portion to be sampled. For heaps it is necessary to specify the points of entry of the sampling trier and the directions of insertion, so that in theory each particle in the heap has an equal chance of being included in the incremental sample. For materials on the move via a conveyor belt, or during unloading operations, it may be convenient to sample on a time basis providing that the flow rate is reasonably constant.

The number and size of incremental samples having been established will also define the size of the aggregate sample. The next part of the plan will discuss how the incremental samples are combined to form the aggregate sample and whether any further treatment is required, *e.g.* reduction in size (reduced sample). The final part of the plan will describe the number of laboratory samples to be produced,

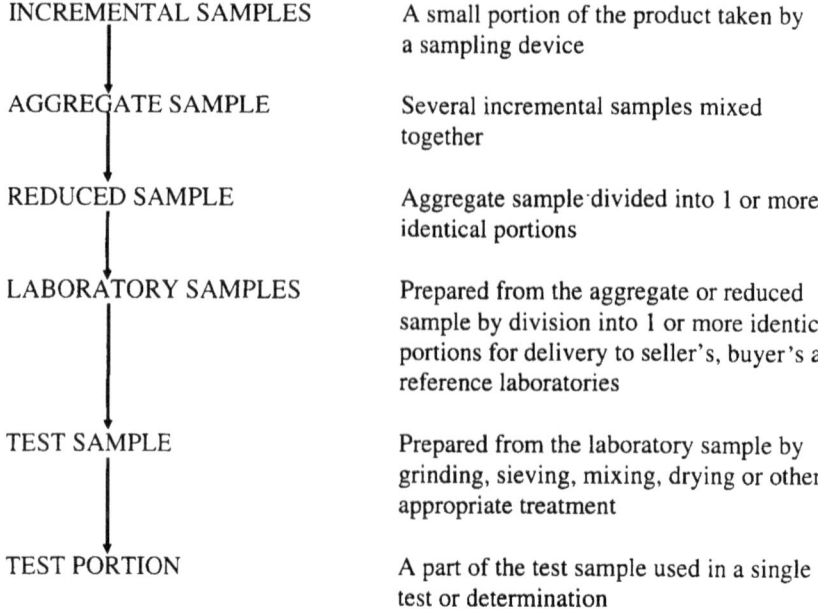

INCREMENTAL SAMPLES	A small portion of the product taken by a sampling device
AGGREGATE SAMPLE	Several incremental samples mixed together
REDUCED SAMPLE	Aggregate sample divided into 1 or more identical portions
LABORATORY SAMPLES	Prepared from the aggregate or reduced sample by division into 1 or more identical portions for delivery to seller's, buyer's and reference laboratories
TEST SAMPLE	Prepared from the laboratory sample by grinding, sieving, mixing, drying or other appropriate treatment
TEST PORTION	A part of the test sample used in a single test or determination

Note: Care should be taken to ensure that representativity is maintained during mixing and sub-sampling operations. Such steps should be kept to a minimum.

Figure 3 *Hierarchy of sampling terms*

the nature of the containers for sample storage, any preservative required, as well as instructions for treatment on arrival at the laboratory and storage before analysis commences. A hierarchy of sampling terms is shown in Figure 3.

This part of the sampling plan will answer the questions:

- How many incremental samples are to be taken?
- How big will each incremental sample be?
- How will the samples be obtained in terms of equipment used, position of sampling, time of sampling?

2.4 Other Considerations

In addition to the above parameters it may be necessary to take account of legal and safety aspects. Such factors can only be discussed in general terms because they are specific to the particular sampling problem.

2.5 Sampling Checklist

A checklist is appended summarising the factors to be considered when drawing up a sampling plan. It should be possible to use this in a wide variety of sampling operations. In all cases detailed written instructions must be provided. These should be quantitative wherever possible.

Finally, where sampling is envisaged on a continuing basis, it is prudent to review the sampling plan from time to time in the light of experience and the results obtained to see whether the plan can be improved.

Checklist for sampling plans (non-environmental)

1. Define the objectives of the programme and obtain agreement from other partners.
2. What will happen to the results produced?
3. What decisions will be taken on the basis of the results obtained? What is the 'value at risk' (in cash terms) of getting the final result wrong by 0.1%, 1%, or 10%?
4. Is the uncertainty resulting from sampling and analysis activities known and of a magnitude that permits the decisions envisaged in 3 to be taken on the basis of sound data? What is the value at risk (in cash terms) of getting the final result wrong by 0.1%, 1%, or 10%.
5. What analytes (or constituents) of the material are to be determined?
6. Which analytical methods are to be used?
7. Have the accuracy and precision data for the method(s) been established?
8. Has the material to be sampled been properly characterised in terms of description, location, batch number, size and homogeneity, *etc.* as appropriate? Is the material to be sampled homogeneous and from a static heap/container, or sampled whilst in motion?
9. If the material is in discrete packages, *e.g.* sacks, bags, drums, *etc.*, how many of these packages are to be sampled and how will they be selected in a random manner?
10. How many incremental samples are to be taken and in what manner?
11. What is the total volume (mass) of the aggregate sample?
12. Will the aggregate sample be reduced in size?
13. What is the size of the final laboratory sample? Are more than one aliquot portions to be prepared?
14. Describe the sample containers and arrangements for transport/storage to the laboratory. Are any preservatives to be used?
15. Describe any sample pretreatment operations, *e.g.* grinding, riffling, moisture determination, *etc.*
16. What arrangements have been made to review the sampling plan?

3 Different Approaches to Sampling

The way in which the sampling plan is devised and developed depends on a number of factors such as the size of the consignment, the characteristics of the product, access to the product for sampling, location and also on the sampling objectives. Sampling plans may specify that samples are to be obtained by (a)

Random, (b) Systematic, (c) Stratified, (d) Sequential, or (e) *ad hoc* 'snap' means. Some comments on these different approaches are now presented.

3.1 Random Sampling

'Random' in sampling terminology means *without bias*, not haphazard. Increments must be taken in such a way that any portion of the bulk material has the same probability of being included in the sample. Random sampling is generally used when there is very little information available about the material under test, or in the sampling of manufactured products, or in sampling for physical properties, *e.g.* particle size.

3.1.1 Stockpiles of cereals

Increments are taken from the surface as well as the interior of the pile to obtain a truly random sample. However, large heaps can never be satisfactorily sampled in this way if the product is heterogeneous as it is not possible to ensure that each particle has an equal chance of being included in the sample.

3.1.2 Compact solids (metal ingots, concrete)

Random drilling is employed to break down the sample and all the drillings collected to obtain a representative portion.

3.1.3 Manufactured products for quality control

Use of a table of random numbers (or computer generated) is frequently employed to ensure that truly random samples are collected from the batch or consignment. This may be from a line of drums or from packages on a conveyor belt. A simple example to illustrate this point is shown in Figure 4. A single unit to be sampled has been divided into 24 imaginary sections, eight at the top, eight in the middle, and eight at the bottom. Each section is given a number and the numbers of the sections to be sampled are obtained from a set of random numbers. Assuming that

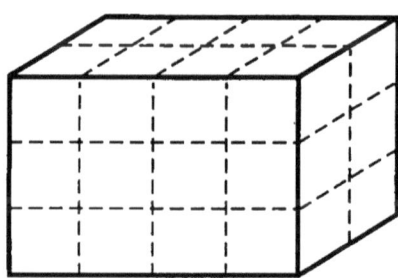

Figure 4 *Division of a single sampling unit*

the increments are to be mixed, then it is important to ensure that the aggregate sample contains an equal quantity of material from each layer. The size of the incremental sample will depend on the nature and heterogeneity of the product. For single components 100 g may be sufficient, whereas for heterogeneous blended products as much as 500 g may be required.

3.2 Systematic Sampling

This is the most commonly used sampling technique. The increments of the sample material are collected at predetermined intervals as defined in the sampling plan.

Some examples of when systematic sampling may be used are described below.

3.2.1 Solid material in motion (on a conveyor belt).

The sample can be taken automatically by periodically transferring a portion of flowing material to a sample container. Samples must be removed from the whole width and depth of the belt as separation occurs with particles of different sizes and densities. This can be achieved either by using automatic equipment or by hand held sampling cups.

3.2.2 Liquids

If the liquid contains suspended solids and segregation is apparent then a systematic plan can be devised to ensure that the sample taken at various depths includes segregated products in the correct proportions. Alternatively it may be better to sample after mixing has occurred, *e.g.* during discharge from the tank.

3.2.3 Manufactured products

More items are collected at problematic times such as just before and after a shift or break, since this is when loss of quality control is most likely to occur. A lower level of sampling is adopted in between problematic periods.

3.3 Stratified Sampling

This is an extension of systematic sampling and involves the division of a consignment into groups (strata). Within each group, material is sampled in proportion to its weight (or volume) as appropriate.

Some examples when stratified sampling may be used are described below.

3.3.1 Wagon load of scrap metals

Sampling is more informative if the components are first sorted into types such as iron, copper, aluminium, and non-metals and then proportional weights can be taken for sampling.

3.3.2 *Material lots delivered at different times*

Proportional weights of materials from each consignment are taken for sampling.

3.3.3 *Sedimented liquid*

The liquid is decanted and weighed if practicable; then proportional weights of liquid and sediment are taken for sampling. In other cases it may be satisfactory to portion the sample on the basis of volume, or depth for a container of constant cross section.

3.3.4 *Rivers, lakes, or reservoirs*

It will usually be necessary to take samples at various depths, including the bottom sediment and at different sites (positions) in the water course.

3.4 Sequential Sampling

This is used to reduce effort for sampling to a particular specification. Samples are taken at predetermined but random intervals checked to specification to establish whether the results are well within or outside the specification. If the result is close to the specification, then more exhaustive sampling and testing procedures may be carried out to pass or fail the product.

3.5 Sampling for Food Surveillance Studies

Surveillance programmes usually of a nation's food supply are undertaken in order to assess intake of nutrients or exposure to additives or contaminants. The data produced have then to be assessed by nutritionalists and/or toxicologists. The programme can be designed from the point of view of the diet of the general population, or that of special interest groups, *e.g.* babies, the very old, those with particular dietary problems, vegetarians, ethnic groups, or diabetics. Hence, the objectives should make clear whether it is designed to monitor exposure of a food component by average, extreme, or special population groups.

Recommendations for the organisation of a total diet study have been made.[5] Samples are purchased at stated intervals throughout the year from different regions of the country and analysed by an agreed protocol. The aim of such studies is to evaluate *changes* in the intake of various chemicals, *e.g.* pesticides, over a period of time. Hence, it is important not to make too many changes to the sampling or analytical protocols unless absolutely necessary so that any observed trends in the data obtained are real and not caused by changes in methodology.

The starting point in any surveillance exercise is to specify the components of the food, *e.g.* nutrients, additives, or contaminants, of interest. Following on from this it is then necessary to identify the foods likely to contain such chemicals based on information available at the time. It might also be necessary to carry out a limited check on other foods, particularly if these constitute a major part of the diet, to confirm that the target chemicals are not present. The number of samples to be taken should then be determined taking into account possible seasonal variations. The place of purchase may also be important, *e.g.* wholesalers, retail outlets, supermarkets, small corner shops, health food stores, or catering establishments (hospitals, restaurants, prisons, schools). Any pre-treatment of the samples, *e.g.* removal of outer leaves, peeling, washing, cooking and storage before analysis will need to be written down in clear instructions. Laboratories which carry out the analysis will have to be supplied with the methods to be used and checks will be needed to ensure that they all perform to an acceptable standard. Finally, consideration for the interpretation of the results should form an important part of the planning process.

Checklist for surveillance sampling for foods

This is to be used in conjunction with the Sampling – General Checklist in Section 2.5.

1. What chemical(s) are of interest?
2. Which foods are likely to contain such chemicals?
3. How many samples of each food are required?
4. Where will the samples be obtained?
5. Have seasonal variations been taken into account?
6. Have special diets/sectors of the population been considered?
7. Should any foods unlikely to contain the chemical(s) be examined?
8. What pre-treatment of the food is required?
9. Which laboratories will carry out the analysis?
10. Have the methods been agreed?
11. How will the performance of the laboratories be controlled?
12. Who will co-ordinate the collection of the samples, transport them to the laboratories, and collect and assess the results?
13. Who will interpret the data, and how?

3.6 Environmental Sampling

The composition of water or the atmosphere will vary with time and with the point of sampling. These aspects must be taken care of when designing the sampling plan. Is the objective to determine fluctuations in analyte concentration over a whole day or from day to day and from place to place, or is the sole objective to obtain an average value? The former objective can best be achieved by a system of continuous monitoring or by the taking of discrete samples at regular intervals from different sampling points. In the latter case for liquids, best practice may indicate the provision of discrete samples which are then mixed together to form a composite sample which needs to be analysed only once. For gases, sampling can be effected by drawing the atmosphere through absorption tubes (or other systems) over a given period of time so that the total pollution load can be calculated following analysis.

3.7 Composite Sample Preparation

In some situations large numbers of samples have to be taken. Analysis of each individual sample separately could prove inordinately expensive, particularly in trace analysis where presumptive 'positives' have to be confirmed by expensive analytical techniques such as gas chromatography–mass spectrometry. An alternative approach which may be appropriate in some cases is to combine equal and representative portions taken from a number of samples to form a single sample for analysis.

Composite sample preparation is not a sampling technique; it is a preparatory technique after the sample has been taken and implemented after agreement with supplier and customer.

4 Safety

Sampling by its very nature can be a hazardous operation. The problems encountered will vary widely depending on the nature of the product to be sampled, the equipment used, and the location. Frequently, the most junior and inexperienced staff are sent to obtain samples. Clearly, this is inappropriate where there is any possibility of hazards arising from the work and all staff should receive appropriate training before commencement of a sampling programme. As the problems are so varied, only general guidelines and recommendations can be offered in this report.

Any activity involves an element of risk. Hence the object of any safety assessment should be to reduce the risks to an acceptable minimum whilst still enabling the work to be carried out. The safety assessment must include (*a*) hazards resulting from the nature of the product to be sampled, (*b*) hazards arising from any equipment used, and (*c*) hazards on site.

Under the Health and Safety at Work Act 1974 both Employers and Employees have duties and responsibilities:

> It is the employer's responsibility to ensure, so far as is reasonably practicable, the health, safety, and welfare at work of his employees. This means that a full assessment of the risks must be undertaken and accordingly the sampling operations used must be safe. Suitable protective equipment necessary for personal safety must be provided.

It is the duty of every employee whilst at work to take reasonable care for the health and safety of himself and of other personnel who may be affected by his acts or omissions at work. Where safety rules have been issued by the employer, it is the duty of the employee to obey these rules.

4.1 Hazards on Site

The sampling officer should have safe access to and from the place where the sample is to be taken under normal and emergency situations. The site should have adequate lighting and ventilation. A full assessment of the risks should be made and included in the sampling plan (Section 2). This should include dangers from nearby machinery, naked lights, smoking, and the need to use non-sparking electrical equipment.

Solids are bulk packed in all sorts of sizes of containers from 25 kg to half a tonne. Such packets are too heavy to lift unaided. Reasonable care must be exercised to prevent puncture by sharp objects, breakage or rupture of the container, and possible discharge of the product, thereby creating a hazard.

Sampling from unsafe sites should be avoided.

4.2 Hazards Arising from the Product

A full COSHH assessment should be made taking into account the physical, chemical, and biological properties of the materials being sampled. This must be included in the sampling plan. Safety Data Sheets must be consulted before starting work. In particular solids may become self-heating in bulk and may ignite. Powders may become airborne and ingested. Dusts can give rise to explosions produced by electrical discharge. The personnel involved in sampling need to be monitored regularly to assess the exposure, particularly when dealing with products of a hazardous nature.

Liquids and gases by virtue of their higher vapour pressure present additional hazards. They are often flammable and can be toxic. Suitable precautions must be included in the sampling plan and adhered to by all sampling personnel.

Sampling of hot discharges, substances under pressure, and material contaminated with radioactive waste requires special systems and equipment.

4.3 Protective Clothing

The sampling plan should specify what protective clothing is to be worn and the equipment to be used. Protective clothing should be used as a 'last line of defence' rather than a safeguard against all possible risks. Too much protective clothing can restrict operations and movement and so create additional hazards. A minimum level of protection should be specified. This may include face masks (or breathing apparatus), overalls, strong footwear, and gloves. Suitable eye protection will always be required. Care must be taken to ensure that the use of protective clothing, cosmetics, or hand (barrier) creams does not contaminate the sample. Some materials are readily permeated by organic solvents which may then constitute a hazard and may also contaminate subsequent samples.

4.4 Safety Checklist

1. Does the sampling plan contain detailed recommendations for safe working practices?
2. Has a full safety assessment been carried out covering the site, the nature of the product, and the safety of personnel?
3. Has adequate protective clothing been made available for the sampling officer?

4. Have the sampling officers been adequately trained?
5. If mechanical or electrical equipment is used for sampling, does it protect the sampling officer from flying fragments and can the machinery be stopped in case of an emergency?
6. In isolated areas, *e.g.* warehouses, where large numbers of samples are to be taken, it may be necessary for safety reasons to deploy two or more sampling officers.

5 Practical Illustrations of Sampling

The application of the general principles of sampling to a number of practical sampling situations will now be described. This will include:

5.1 Sampling from a large heap
5.2 Sampling from packages/sacks/cans/bags
5.3 Sampling from drums
5.4 Sampling from a reservoir or river
5.5 Atmospheric sampling

It should be relatively easy to adapt the suggestions that follow to other sampling problems once the *objectives* have been clearly defined. A general checklist of aspects to be considered when drawing up and implementing a sampling programme is reproduced below. This general plan is modified in the examples that follow and can also be modified in other ways to suit the job in hand.

Sampling – general checklist

1. Has a *SAMPLING PLAN* been drawn up and agreed?
2. Have all aspects of *SAFETY* been considered
    (a) relating to the site where sampling is to be undertaken
    (b) relating to hazards arising from the product?
3. Has the necessary safety *EQUIPMENT* been obtained and checked?
4. Is *legal permission* required for access to the site? Has it been obtained?
5. What is to be sampled?
6. How many samples (increments) are to be drawn (taken)? From where? How? Is any prior mixing required?
7. For contractual (or other) purposes, should the sampling be carried out in the presence of witnesses?
8. If packets are to be selected at random for sampling, how is this randomisation to be arranged?
9. Is the sampling to be carried out 'at rest' or from a bulk material 'on the move'?
10. What arrangements have been made to combine and mix incremental samples and subsequent sub-division into portions?
11. Label the sample container(s) with product description, sampling officer's

name, place, time and date of sampling, unique identification number. Seal the container.
12. Transport to the laboratory (test house) without delay.

This section should be read in conjunction with other parts of the book, especially Section 3, Different approaches to sampling, Section 4, Safety, and Section 6, Equipment for Sampling.

5.1 Sampling from a Large Heap

A large delivery of a commodity may arrive in railway wagons or in a series of lorries. In either case the consignment will probably consist of several lots (or batches) and will end up in a large heap. Large static heaps of a heterogeneous product cannot be sampled satisfactorily. Where the particles of the product differ in size or density, segregation will occur. In static sampling, the tool used would only reach a small fraction of the heap and so the basic premise that every particle must have an equal chance of being included in or rejected from the sample taken cannot be achieved. In such cases, satisfactory sampling can only be achieved during transfer of the heap by conveyor belt from one point to another or during loading/unloading.

If the loading/unloading operation is performed using grabs or shovels it may be desirable to treat each separate grab as a defined lot. These could be numbered sequentially and the numbers selected for sampling by a random process. Where loading/unloading is effected by conveyor belt at constant speed, the sampling units (lots) could be identified by time and, again, a random selection made for taking samples.

Sampling from moving belts can be dangerous. Automatic equipment may be available but should be checked for bias. Specially designed sampling cups are available for taking incremental samples from a moving, falling stream. The cup is passed slowly across the whole width of the stream. This approach is unsuitable if the stream is moving very quickly.

It is also possible to take samples from a conveyor belt by periodically stopping the belt. The sample is then removed by taking a portion of the whole width of the belt in a direction at right angles to the direction of flow. Inevitably this process interrupts the loading/unloading process but is useful as a reference method to check the accuracy (bias) of automatic or mechanical sampling techniques.

In the case of fertilisers, international standards have been prepared.[6]

Where there is no alternative to sampling from a static heap and the heap is not too large (*i.e* less than a tonne) the following notes may prove helpful in drawing up a sampling plan and its subsequent implementation. The unit to be sampled must first be defined. This may simply be the whole heap, if not too large. Otherwise it may be desirable to split the heap into a series of say five separate heaps and to obtain a separate sample from each (imaginary) heap. This approach will also provide data on the variation in composition between different parts of the heap. Then the number of incremental samples required will need to be agreed. Some examples from the UK Fertilisers (Sampling and Analysis)

Table 1 *Loose fertilisers*

Size of sampled portion (tonnes)	Number of incremental samples required
Up to 2.5	Not less than 7
>9 to 11	Not less than 14
>20 to 22	Not less than 21
>48 to 51	Not less than 32
>76	Not less than 40

These figures are calculated from the formula:
Size in tonne(s) $\times$ 20 = 20s
Increments required = $\sqrt{20s}$, raised to the nearest whole number.

Regulations 1991[7] are shown in Table 1.

For small heaps (say 5 tonnes and below) it may be prudent to take a larger number of incremental samples than is obtained by using this formula. The minimum number of incremental samples could be taken as 10 in such cases even if the size of the heap is below 5 tonnes.

The size of the incremental sample will also have to be agreed. It may fall within the range 100 g to 1 kg depending on the equipment used and the product. The incremental samples then need to be mixed to form a composite or aggregate sample which in turn may have to be reduced to form the final sample. This may have to be split into aliquot portions if the sample is to be examined by more than one laboratory. Normally, the final sample should be four times the size of the laboratory sample. The sequence of operations is illustrated in Figure 5.

The final/laboratory samples must then be packed in clean containers for transport without delay to the laboratory. The container should not contaminate the sample or remove trace constituents from the sample by adsorption. It is good practice to choose a container of a size such that it is almost completely filled by the sample. This reduces the effect of possible oxidation or change in the physical characteristics of the sample by shaking during transport to the laboratory. This could be crucial if particle size is to be determined. If the container cannot be completely filled by the sample and physical tests are required, it may be possible to fill the space above the sample by an inert material, *e.g.* crushed paper. The container must be sealed and labelled and any necessary paperwork (sampling report) completed.

Sampling from a heap: checklist

1. Read the *SAMPLING PLAN* and any supplementary instructions.
2. Has *legal permission* to take samples been obtained?
3. Is any special safety clothing/equipment required?
4. Is all sampling equipment/containers clean and dry?
5. Is the heap to be sampled clearly defined and identified?
6. Where and how will incremental samples be drawn from the heap?
7. Is any pre-treatment of the heap required, *e.g.* mixing, moving?
8. How many incremental samples are required?

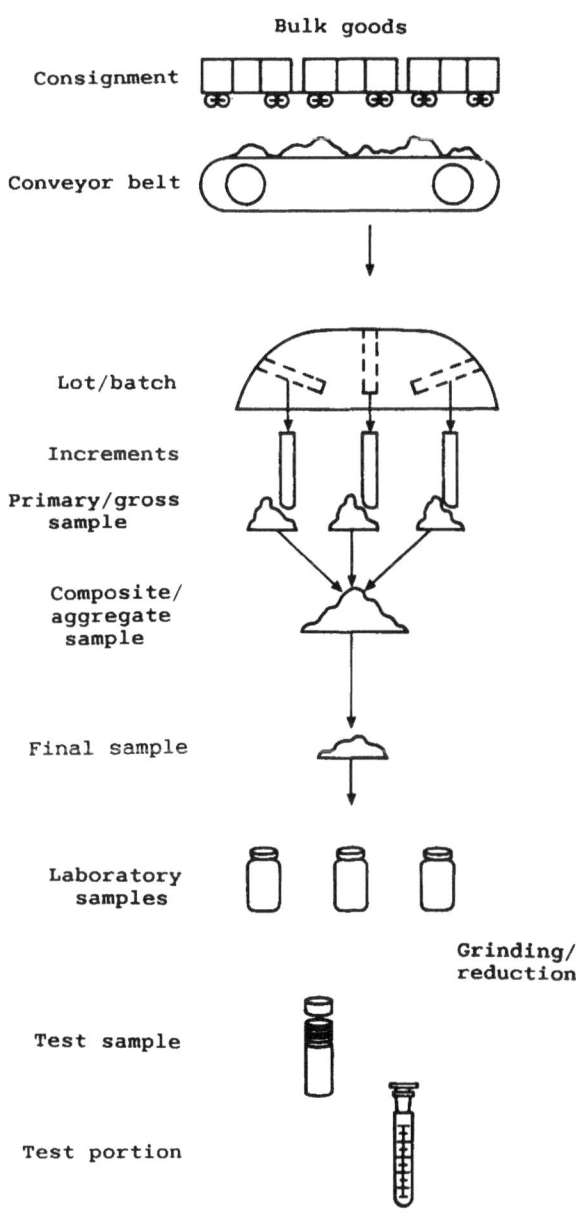

Figure 5 *Sampling scheme for bulk goods*

9. What is the approximate weight required of all incremental samples?
10. How are the incremental samples to be combined to form the aggregate sample?
11. Is the aggregate sample to be sub-divided into one or more individual test samples, and how?
12. Label the sample container(s) with product description, sampling officer's name, time and date of sampling, unique identification number. Seal the container.
13. Arrange transport to the laboratory without delay.
14. Complete any necessary paperwork/documentation.

5.2 Sampling from Packages/Sacks/Cans/Bags

One advantage of packaged goods over a loose heap is that in the former case the boundaries of the product are clearly defined. However, the product may be in a lorry when only a limited number of packages would be accessible for sampling. Alternatively, even when stacked on pallets in a warehouse, again only the packets on the surface could be readily removed without the co-operation of the warehouse staff or by a great deal of physical effort. Again sampling operations may be easier during loading or unloading.

The sampling plan must first identify the global number of packets which constitute the consignment, delivery, or portion to be sampled. It is then necessary to state how many of these units are to be sampled. In many cases the formula used is:

Number of packages	*Minimum number to be sampled*
100 or less	Not less than 10
Above 100	$\sqrt{}$ (number present), rounded up

Having decided how many packages are to be sampled, it is then necessary to decide how these packages are to be selected from the bulk consignment in a random manner so that each package in the lot has an equal chance of being selected for sampling.

Where each package weighs less than 5 kg, the entire contents may be taken as the incremental sample. For packages above this weight it will be necessary to empty the contents out onto a clean surface and mix before sampling. A portion of defined weight can then be removed by shovel, sampling spear, or other device as appropriate or as specified in the sampling plan. In all cases the dimensions of the sampling equipment are critical. The incremental samples so obtained then have to be mixed together to form the aggregate sample, which should be of the size specified in the sampling plan. It may then be necessary to reduce the size of the aggregate sample by riffling, or coning and quartering, as described in Section 6. Final laboratory sample(s) must then be prepared and transmitted to the laboratory for testing without delay. Suitable containers must be used which are clean and dry and of a material which does not either contaminate the sample or remove constituents by adsorption. The container must be suitably sealed and labelled.

The sequential operations involved in sampling from packaged products are illustrated in Figure 6. A checklist containing the most important points to bear in mind when visiting a site to commence sampling follows.

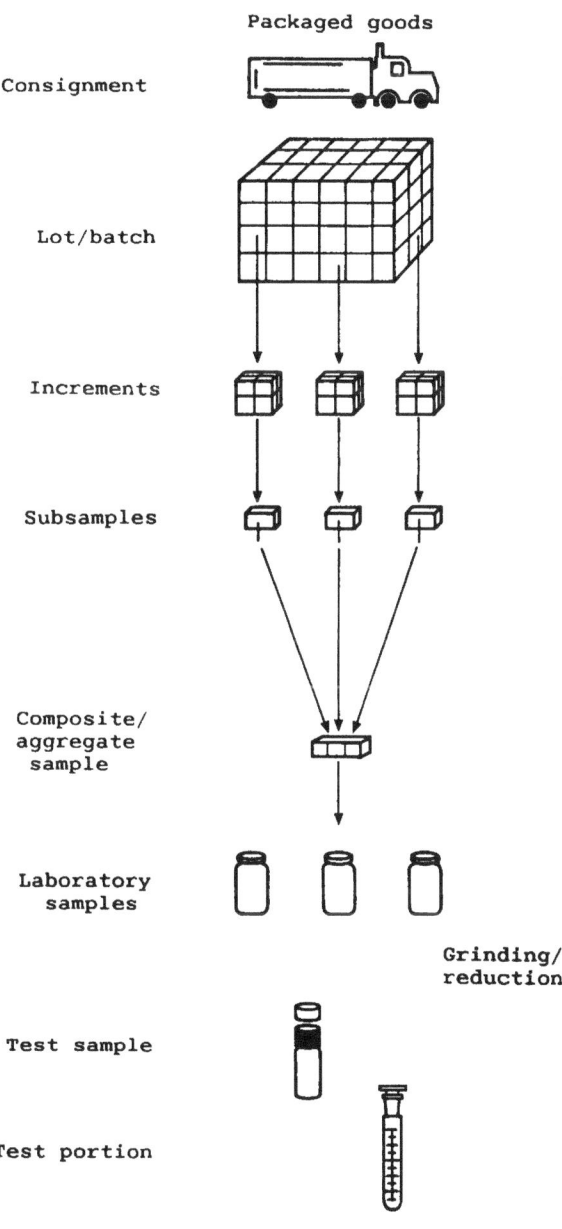

Figure 6 *Sampling scheme for packaged goods*

Sampling from packages/sacks: checklist

1. Read the *SAMPLING PLAN* and any supplementary instructions.
2. Has *permission* to take samples been obtained?
3. Is all sampling equipment, including containers, clean and dry?
4. Can the number of packages/sacks be clearly identified?
5. How many are to be sampled? How will they be selected in a random manner?
6. How many samples are to be drawn from each package/sack?
7. Ensure that the removal of any package does not cause instability in the remainder of the packages.
8. Should the packages/sacks be emptied and/or mixed, prior to taking samples?
9. Are the samples to be drawn from different layers or locations in the package/sack?
10. How are the incremental samples to be mixed to form an aggregate sample?
11. Is the aggregate sample to be sub-divided into one or more test samples?
12. Label the sample container(s) with product description, sampling officer's name, place, time and date of sampling, unique identification number. Seal the container.
13. Arrange transport to the laboratory without delay.
14. Complete any necessary paperwork/documentation.

5.3 Sampling from Drums

A consignment or delivery may consist of a large number of drums which may contain a product prepared from one or several batches. The total number of drums will be known and the sampling plan should indicate how to calculate the number to be selected for sampling and how this selection process is to be achieved so that each drum has an equal chance of being selected. As for packages, the number selected for sampling is often the square root of the total number in the consignment. The drums can be numbered from 1 to *n* and the selection process completed using a table of random numbers or a calculator with a random number generation facility. Where the total number of drums is less than (say) 6 it will probably be necessary to take a sample from each drum in the consignment. If the drums contain a liquid, they may need to be rolled or the contents otherwise mixed before sampling, particularly if suspended matter is present. Alternatively, it may be more convenient to sample during emptying. This is particularly true for very large containers where portions of around 500 ml should be collected at regular intervals during discharge. Where the contents of the drum are not more than (say) 100 litres approximately equal volumes can be removed from each drum and combined to form the aggregate sample. For small drums (say 1 litre or less) the entire contents should be taken as the incremental sample.

Drums which have been stored outside in cold conditions may contain liquid and crystalline constituents. Such segregation may not always be clearly visible to the sampling officer and will depend on the properties of the product. Ideally, the drums should be carefully warmed until only a single phase is present before sampling. Drums containing viscous fluids may also be easier to sample if slightly warmed.

Drums containing a solid can be sampled by probes if the product is not too inhomogeneous. If in doubt, check the variation in composition from successive incremental samples. Samples must be drawn from different depths of the drum and the probe inserted at different angles to the surface. Precautions should be taken to avoid loss of volatiles.

The incremental samples are then mixed together to form the aggregate sample. The final and laboratory samples are obtained by withdrawal of a portion of suitable size as specified in the sampling plan. The containers must be airtight, especially for liquids, and filled as much as possible leaving the minimum void space consistent with prevention of breakage by possible thermal expansion during transport. After sealing, the containers should be labelled and transported to the laboratory without delay.

Caution must be exercised when opening drums or sealed containers which have been exposed to temperatures higher than ambient, or following melting. Under such conditions high pressures may build up within the containers. The opening of closures must be undertaken with great caution, allowing excess pressure to vent fully before finally opening. Such venting procedures are best effected by covering the closure with paper towels or bags preventing spraying of contents. Under conditions of extreme cold partial vacuum may exist inside containers. In the case of drums stored outside, ensure that standing water or other contaminants are not drawn into the drum on opening.

A checklist of the main operations involved in the sampling of drums is produced below.

Sampling from drums: checklist

1. Read the *SAMPLING PLAN* and any supplementary instructions.
2. Has *permission* to take samples been obtained?
3. Is any special safety clothing/equipment required?
4. Is all sampling equipment/containers clean and dry?
5. If more than one drum is to be sampled, is the consignment numbered in a logical/consistent manner?
6. How many of the total number of drums are to be sampled?
7. How are these drums to be selected in a random manner?
8. Are the drums to be rolled (or otherwise treated) before sampling?
9. How many samples are to be drawn from each drum?
10. Are the samples to be drawn from different layers or locations in the drum?
11. If the drums contain a liquid, is there likely to be any sediment or suspended matter? Should the sampling operation be undertaken whilst the liquid is in transit from the drum to another container?
12. How are the incremental samples (solids or liquids) to be mixed to form the aggregate sample?
13. Is the aggregate sample to be sub-divided into one or more test samples?

14. Label the sample container(s) with product description, sampling officer's name, place, time and date of sampling, and unique identification number. Seal the container.
15. Arrange transport to the laboratory without delay.
16. Complete any necessary paperwork/documentation.

5.4 Sampling from a River, Lake, or Reservoir

The previous examples have involved static conditions where the composition of the consignment does not change with time or position in space (heap or container). Other sampling situations involve dynamic systems where the composition of the product is changing with time and with the point of sampling. Such situations are encountered in environmental and clinical chemistry and during manufacturing process monitoring.

The sampling plan is crucial to this type of operation and in most cases the plan will need to be very detailed. The location of sampling is important and will depend on the purpose of the exercise and on other environmental factors such as the season, weather conditions and the presence of outflows or weirs. Consideration will need to be given to surface or depth sampling, the effects of sediment and flora that may be present, water flow rates and the technique used to obtain the samples. Safety considerations are paramount and will vary from site to site. **Sampling should not be attempted alone.** Special equipment, *e.g.* boats, may be needed for deeper waters. Currents and underwater objects may present additional hazards.

Reservoirs and rivers are usually in remote areas, often at high altitudes. In winter conditions consideration should be given to the possibility of sudden adverse conditions such as snowfall. Basic survival equipment would be a useful precaution. Reservoirs and river banks can be very treacherous places with uncertain footings adjacent to deep water. Under such circumstances the wearing of flotation equipment is necessary. Consideration should be given to the availability of the back-up personnel, and access to the site, for normal and emergency use, should be identified. The practice of sampling officers working together for mutual safety is essential. The use of safety harness and ropes may be required.

In rivers, extra care will be needed when working in deep water, fast moving currents, or close to a weir or outfall. Safety aspects of sampling in natural water, potable water, treatment plant,[8] sewage, and effluent [9,10] have been detailed. Care must be taken to ensure that when wading the mud or sediment disturbed does not affect the sample collected. Samples collected close to a bank in stagnant water may not be typical of the fast flowing centre of the stream. Surface water may differ from layers close to the bottom where constituents are slowly leached out of the mud or sediment.

The need for snap or sequential sampling will have to be considered and form part of the agreed sampling plan. Times and frequency of sampling must be clearly specified. Alternatively, continuous monitoring may be a better option if only simple analytical determinations are to be made. In this case the location of the monitoring sensors and security of any logging equipment used needs to be considered. Sampling for benthic macro-invertebrates requires special equipment and techniques. Sampling of surface films can be a problem as the thickness of the film may be only one molecule and certainly will be small in comparison with the depth of the water layer. Oil spillages suffer rapid evaporation of the lighter fractions which eventually may lead to an increase in specific gravity so that the residue instead of floating then sinks. Photodegradation may also occur. Such processes are also influenced by wave action which can produce a product with a consistency of slurry (or sludge) from an oil spillage.

The container used to transport the sample is important when testing for trace constituents in particular. Small amounts of sodium or boron may be leached out of new glassware. Equally, trace constituents, *e.g.* metals or polynuclear aromatic hydrocarbons, can be adsorbed onto active surfaces. Dark coloured bottles should be used for light-sensitive constituents of interest. Polythene or PTFE vessels are often used for water when trace elements are to be determined. Glass containers are preferred with PTFE lined bottle caps for trace organics. If glass containers are used when inorganic components are to be measured, some mineral acid should be added as preservative. This should be stated on the label and in the accompanying paperwork and should form part of the sampling plan. It may be necessary to filter some samples either on site or on arrival at the laboratory, and the filtration conditions need to be specified. Sample containers should be completely filled to exclude air which may cause oxidation of biological and chemical components.

The cleaning agents used for the sample containers also need to be considered, *e.g.* use of chromic acid for trace organics and phosphate-free detergents for phosphate determination. Certain samples may need to be chilled during transport to prevent degradation of components; chilled boxes (ice-packed) or cabinets can be used.

A checklist of the most important factors is given below.

Sampling from a reservoir or river: checklist

1. Read the *SAMPLING PLAN* and any supplementary instructions.
2. Has *legal permission* to take samples been obtained?
3. What special safety equipment and clothing are needed? Sampling from lakes or deep rivers will require a boat and life jacket, and the means of anchoring the boat on station. Waders will be required for shallower rivers.
4. Sampling from banks, bridges, or weirs may be preferable. The site must be selected carefully (see 1).
5. Surface or depth sampling required? Flowing systems may vary from middle to edge and from site to site. The position of outflows may be critical.
6. The effects of currents and the presence of underwater objects (rocks, mud, debris) must be borne in mind at all times.
7. Discrete or sequential sampling taken at constant increments of time or flow may be required (see 1).
8. Are the samples taken to be composited?
9. Is any preservative to be added to the container?

10. Is the sample to be filtered?
11. Record the temperature of the water, rate of flow, and depth as appropriate. Comment on the weather conditions.
12. Label the sample container(s) with a unique number, name of sampling officer, date and time of sampling, location and information in 9, 10, and 11, as appropriate.
13. Arrange transport of samples to the laboratory without delay. Cooling to 4°C may be necessary to slow the activity of micro-organisms.
14. Complete any necessary paperwork/documentation.

5.5 Atmospheric Sampling

Atmospheric sampling is carried out to monitor levels of pollutants in the atmosphere. The validity of any sampling procedure should be tested if the environmental conditions are likely to affect the accuracy of the method. The selection of the sampling system for air pollutants will depend on its purpose. Published methods for the sampling of the particulate matter are available, but validated methods for the sampling of gaseous species are limited. The monitoring of the particulate matter may be made from the particle size distribution, from chemical or physical processes, or from the mass of the particulate matter collected. Occupational hygiene sampling may be adopted to monitor the exposure of personnel in the workplace. A detailed report by the Environmental Health Executive recommends strategies to monitor discharges in particular sites.[11]

Sampling sites can be divided into two categories:

(i)   Confined areas (boreholes, chimney stacks)
(ii) Large internal areas (factory atmospheres) and open atmosphere

5.5.1 Confined areas (boreholes, chimney stacks)

In confined areas, the gases are generally rapidly moving; thus single samples are generally representative and can be assumed to be of the same composition, although changes with time may occur. However, in areas such as boreholes or landfill sites gases are static and may be stratified. In addition the composition may be affected by weather conditions such as rainfall so it may be more appropriate to take several samples at fixed intervals or over a period of time.

Air flows in stacks may be linear or turbulent and there is likely to be stratification of the sample flows across a duct. The sampling strategy is dependent on the size and the shape of the duct, the particulate matter released from the stack, and also the flow and temperature of the gas. Samples are normally taken iso-kinetically by matching the velocity of the sampling stream with that of the gas in the duct. The air flow may vary, in which case more than one sample will be necessary. It is important that the sampler has information relating to the operation of any plants associated with stack sampling so that an appropriate sampling device is adopted.

5.5.2 Large internal areas (factory atmosphere) and open atmospheres

Over large internal areas the sampling is generally used to provide information on background levels. The atmosphere is likely to vary, as the sources of pollution will be distributed unevenly. Further problems are caused by weather conditions, for example wind direction, convection currents, and precipitation in the open.

To overcome these problems simultaneous sampling at a number of locations at fixed or random time intervals can be adopted. This method of sampling is time consuming but provides information regarding the sources and changes in the levels of pollution over a specified period (*e.g.* a normal working day). Details on general methods for sampling airborne gases and vapours are given in Reference 12.

Static sampling is generally used to provide information on background levels in the atmosphere but if the assessment of individual exposure is required personal samples must be taken from the breathing zone of the worker.

Continuous or discontinuous sampling at a single location is used to monitor the levels of pollutants which may cause toxicity. The air sample is collected over a period of time (12–24 h). The minimum period of sampling is normally limited by the pump capacity and the capacity or breakthrough characteristic of the collecting media.

5.5.3 Samplers and detectors

For the direct measurement of gases, portable samplers and/or analysers are available, some of these can also be located in confined spaces, such as in the effluent gas duct. Various in-line devices such as colorimetric gas analysers, infra-red (IR) detectors, and combustible gas indicators can be used for the detection of organic solvent vapours. The IR detectors are suitable for the measurement of flammable gases including methane, ethane, butane, *etc.*

If the air pollutants are present at low levels, the existing analytical techniques may not be sensitive enough for the direct measurement of gases. In these cases, the compound of interest can be trapped either chemically or physically at the time of sampling, thus avoiding the necessity of manipulating large volumes of gases in containers.

5.5.4 Other safety considerations

On sampling locations where there is a danger of falling, the use of safety harness may be necessary. If the air around the sampling location is of a hazardous nature and there is a danger of asphyxiation from fumes then the use of breathing apparatus is essential. When sampling in confined areas, the presence of oxygen should be monitored using 'failsafe' devices which produce an alarm when oxygen levels are deficient and should also give a signal if the device is not functioning. When working at heights, for example on chimneys, the conditions of high winds should be avoided. Under certain conditions unaccompanied sampling may be inadvisable.

Atmospheric sampling: checklist

1. This checklist covers environmental sampling at outside locations as well as sampling internal atmospheres, *e.g.* in factories.
2. Read the *SAMPLING PLAN* and any supplementary instructions.
3. Has *legal permission* to take samples been obtained? Have safety aspects and hazards been fully assessed?
4. The site of sampling is crucial. This includes location, effect of nearby buildings, relationship to wind direction, height above ground, and distance from source of pollution for external sampling. Other atmospheric conditions such the initial and final temperature, pressure, and gas flow rate should be recorded. Similar considerations may be important for internal sampling.
5. Is personal or general atmospheric monitoring to be performed? Record the identity and movements of the person wearing the monitor.
6. Is intermittent or continuous sampling to be practised?
7. What is the duration of sampling, or the volume of sample to be collected?
8. Are you certain that the collection medium will not become exhausted during sampling?
9. Will any operations in the vicinity interfere? Record any such observations.
10. Label the samples with name of sampling officer, date, time and place of sampling, unique sample number, and weather conditions as appropriate.
11. Arrange transport to the laboratory without delay.
12. Complete any necessary paperwork/documentation.

6 Equipment for Sampling

The sampling plan should detail the equipment to be used and the necessary cleaning and storage procedures. Sampling equipment must be such that it will not contaminate the sample from its constituents or through lack of cleanliness. The equipment must be constructed from a suitable material and be of adequate strength suitable for the purpose.

All the sampling tools should always be left fit and clean, for immediate reuse. In addition, surface contaminants on the implements, from storage in the laboratory, or from previous use, may be transferred to the sample. The recommended cleaning and storage procedures for the implements must be followed.

It is also important to ensure that the sampling equipment and the containers for the samples are mutually compatible. All contact components should be dedicated to a particular sampling site in order to minimise the possibility of cross contamination.

The implements that come in contact with the sample are widely recognised as one of the most serious sources of contamination and this possibility must always be borne in mind when interpreting the final results. Most metallic instruments should be avoided when sampling biological materials that are to be analysed for trace elemental contents. Similarly, certain plastics can adsorb charged particles and hence may not be suitable when organic constituents are to be determined. Plasticisers may be leached out into the sample causing contamination.

6.1 Sampling Equipment for Solids

To provide unbiased sampling it is crucial that the whole of the lot of the material is easily accessible to the sampler. This is achieved, as far as possible, by taking a number of separate samples (increments) and then mixing these incremental samples to form an aggregate sample. The samples may be taken manually or with an automatic sampling device.

Sampling probes can be used to remove samples from the centre of the heap, but it is difficult to ensure that both the location and the increments contain the correct proportion of large and small particles. Fewer problems are encountered when the particles being sampled are similar, *e.g.* flour, sugar, or for other products such as cheese and butter. The contents of the sampling tool are then emptied into clean, labelled containers or strong polythene bags and placed in a dry storage environment.

6.2 Manual Sampling

Sampling of solid materials using scoops, shovels, probes, and augers is widely used. However, for large consignments semi-automatic or automatic sampling is preferred as the manual sampling method is slow, laborious, and prone to bias.

A wide range of tools for manual sampling is described in the British Standard[13] and in the literature.[14,15] In some cases particular tools may be specified, but in others the sampler must be guided by experience and use the appropriate apparatus according to the particle size of the product.

6.2.1 Scoops/shovels

The width of the scoop or shovel should be at least 2.5–3 times the largest dimension of the sample particles. Figure 7 shows the design of the scoop recommended by the Japanese Industrial Standards, published in the JIS 8100.[16] Table 2 gives the dimension of the scoop for particles in the range 0.25–125 mm. Particles larger than 150 mm should be crushed before sampling.

The size of scoop used for sampling is governed by the particle size of the material. If the scoop is too small for the particle size of the material then larger particles will roll off, resulting in bias. On the other hand if the scoop is too large an unnecessarily larger sample will be obtained for a given number of increments.

With fine material it is advantageous to take a larger scoop to minimise moisture losses. Table 3 shows the effective mass capacity of certain scoops for material of various bulk densities. The volume of the material taken with larger scoops may be too great; in such cases it may be more convenient to draw imaginary segments across the heap and take entire portions of the increments by a shovel.

Sampling by a shovel is easiest when the sample is moved from one location to

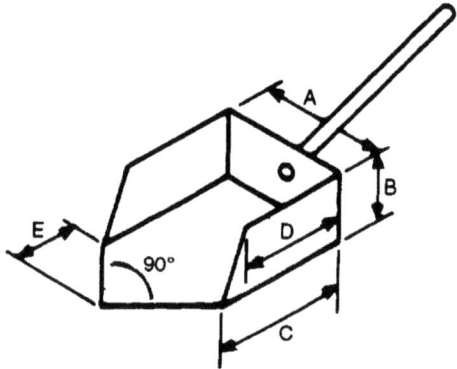

Figure 7 *Sampling scoop for solids: dimensions are given in Tables 2 and 3*

Table 2 *Dimensions of scoop for collecting increments*

Scoop No.	Maximum particle size (mm)	Dimensions of scoop (mm)					Volume (ml)
		a	*b*	*c*	*d*	*e*	
125	125	300	120	300	250	120	10 000
100	100	250	110	250	220	100	7000
75	75	200	100	200	170	80	4000
50	50	150	75	150	130	65	1700
40	40	110	65	110	95	50	790
30	30	90	50	90	80	40	400
20	20	80	45	80	70	35	300
15	15	70	40	70	60	30	200
10	10	60	35	60	50	25	125
5	5	50	30	50	40	20	75
3	3	40	25	40	30	15	40
1	1	30	15	30	25	12	15
0.25	0.25	15	10	5	12	0	2

Table 3 *Effective scoop capacities at different bulk capacities*

Scoop No.	Volume (ml)	Capacity of scoop (g) at following bulk densities			
		0.5 t m^{-3}	1.0 t m^{-3}	2.0 t m^{-3}	3.0 t m^{-3}
150	16 000	8000	16 000	32 000	48 000
50	1700	850	1700	3400	5100
20	300	150	300	600	900
5	75	38	75	150	225
1	15	8	15	30	45

another. The method consists of generally taking every fifth or tenth shovel. The advantage of the shovel sampling is that it can be applied to large lots of material, up to several tonnes. However, it cannot be used if the particles of the material exceed 50 mm in diameter.

6.2.2 Pipes, spears (triers or swords), probes

A pipe, spear, or probe consists of a metal tube which is inserted vertically into the bulk sample to retain a core of the sample of length 40–50 cm in length and 15–25 mm diameter. This sampling procedure is only suitable for fine materials which allow easy access with the tube.

Spears in their simplest form consist of a piece of tubular steel with one end closed to permit easy access into the sample material (Figure 8). The divided spear thief is a tube with a slot running its entire length. The slot is covered during

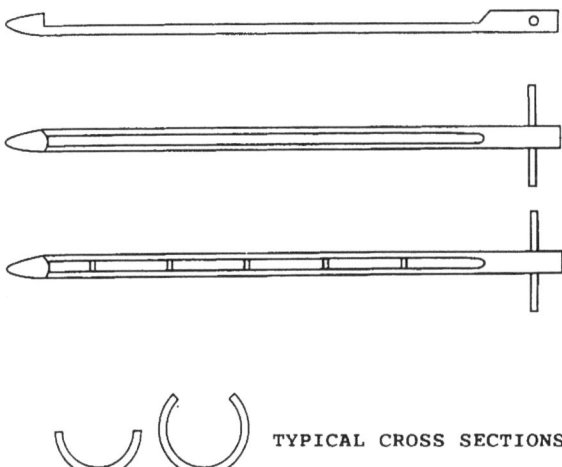

Figure 8 *Open-sided sampling spear and divided spear for dry, free-running powders* (Reproduced by permission from 'The Sampling of Bulk Materials', by R. Smith and G.V. James, The Royal Society of Chemistry, London, 1981)

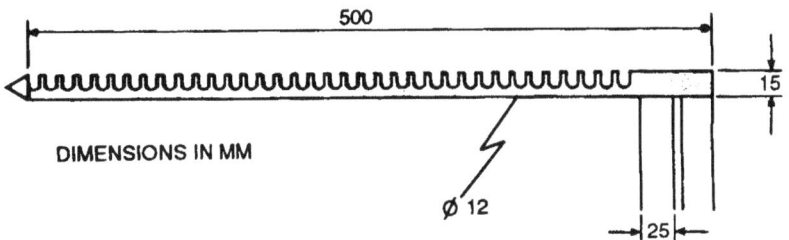

Figure 9 *Shuttered sampling spear for dry, free-running solids* (Reproduced by permission from 'The Sampling of Bulk Materials', by R. Smith and G.V. James, The Royal Society of Chemistry, London, 1981)

insertion of the spear, opened when in position, and closed again before withdrawal. The sample is then removed from the slot.

For very free running solid materials a shuttered spear (Figure 9) may be used. The spear fits inside a tubular sheath. The shuttered spear is inverted into the material and the sheath is rotated by 180° to allow the solid into the open spear. The spear is then closed and withdrawn to collect the sample.

For sampling of soft solids such as cheese, tapered spears are used.

A bag probe can be used to sample from a bag: the probe should be thrust diagonally across the bag to obtain an unbiased sample. For a large bulk material, a bulk probe must be used which is similar to the bag probe but with sufficient length to reach the bottom of the bulk pile.

Grain probes (Figure 10) or double tube compartment probes are used to sample grains, such as rice, beans, seeds, cornmeal, *etc*. The grain spears collect a small sample from considerable depths within the consignment. The spear is pushed into the grain and automatically opened to collect the sample as it is withdrawn. Considerable strength may be required to drive the spear into a large heap of tightly packed particles.

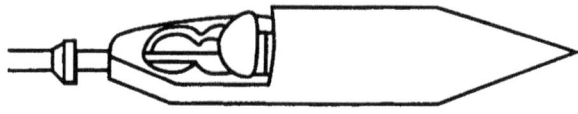

Figure 10 *Grain probe*
(Reproduced by permission from 'The Sampling of Bulk Materials', by R. Smith and G.V. James, The Royal Society of Chemistry, London, 1981)

6.2.3 Augers

Hard materials ranging from natural (rocks, clay) to man-made (concrete, metal ingots) are sampled with a cutting action device to bore through the compact mass with augers (Figure 11). As the auger is rotated, the tube is pushed into the material and all the drillings are collected to take a representative sample.

Drilling machines are commercially available which can be used to take cores of material ranging from 12 to 300 mm in diameter.

This technique can also be applied to frozen packages of fruit in containers (30–50 lb) which are sampled with a corrosion resistant auger 1–1½ inches in diameter and approximately 20 inches long, which can be operated by an electric motor. A lot of heat is generated by augers whilst drilling and this may cause moisture changes or compositional changes in certain samples.

6.3 Semi-automatic Sampling

This is a combination of manual and automatic sampling where the primary sample is collected automatically, and then sub-sampled manually. The advantages are low running and installation costs, automatic collection of increments, and less risk of equipment failure.

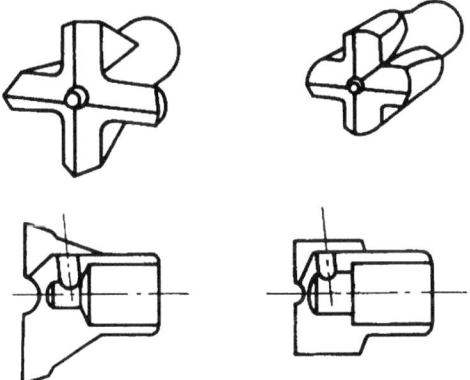

Figure 11 *Cutting tool for sampling drill*
(Reproduced by permission from BS 5309, Part 4, 1976)

6.3.1 Pneumatically driven probe

The pneumatic probe permits a cross section of the material to be sectioned as the probe is inserted from the top to the bottom of the load. All of the sample can be recovered without any changes in the moisture content.

6.3.2 Grab sampler/mechanical digger

Large loads of material, up to several thousand tonnes, are easier to sample at set intervals during loading or unloading, using a grab sampler attached to a crane. The grab sampler can be set to remove samples at specific intervals. If large increments need to be taken then a mechanical digger is more appropriate.

6.4 Sub-sampling – Solids

The process of sub-sampling produces a relatively small representative material from a large bulk of material. At one or more stages, the sample has to be further reduced in weight or sub-sampled. For representative sub-sampling, it may be necessary to reduce the particle size of the material. The procedure and implements used for sub-sampling and to reduce the particle size will depend on various factors such as the nature, the quantity, the particle size of the material, and the required particle size after milling.

6.4.1 Method of coning and quartering

The aggregate sample is placed on a clean flat surface. The sampling surface should not absorb any moisture from the product. Using a shovel, or a suitable tool according to the particle size of the sample, the lot is formed into the shape of a cone. Any fine material remaining is spread at the top of the cone. The top of the cone is then flattened and divided into four roughly equal quarters. One pair of opposite quarters is then removed, the remaining pair is mixed and formed into a separate cone. The process of quartering and rejection is continued until a suitable quantity of the sample remains (1–2 kg for feeds and fertilisers). It may be used for materials up to 50 tonnes in which the particle does not exceed 5 cm in diameter. This sampling procedure is applicable to a wide range of materials and requires few tools.

6.4.2 *Method of riffling*

As an alternative, the gross sample can be reduced by using a mechanical quartering device, such as a sample divider or riffler. Riffle dividers are particularly useful with large samples which are normally more difficult to sub-sample. They are available in many sizes ranging from bench to floor mounted models.

A riffler (Figure 12) can be used to divide a sample into two approximately equal parts. The distance between the slots can vary and should be at least three times the size of the largest particle in the lot. The material to be sub-divided is poured into the top of the box or a feeder and the sample is divided longitudinally and emerges as two equal portions. The procedure of dividing is repeated, discarding the portions from alternate slots, until a portion of suitable size is obtained for analysis.

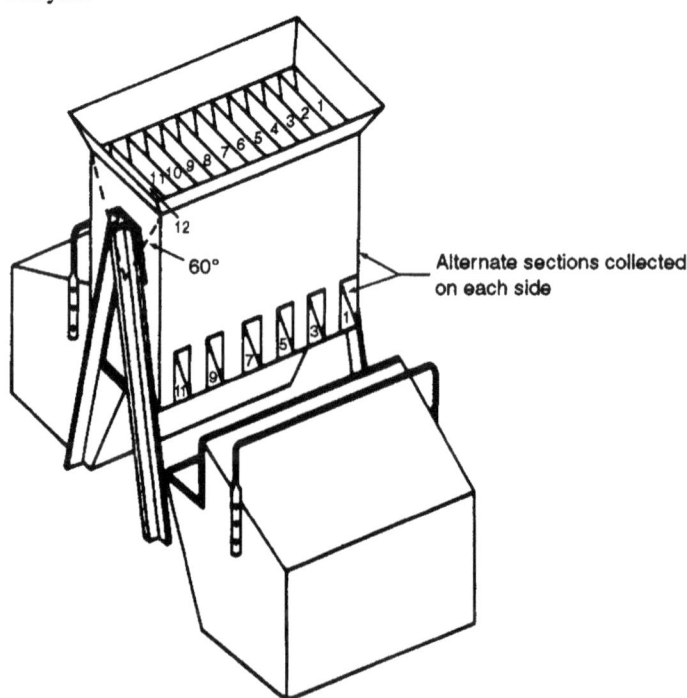

Figure 12 *Riffle sample divider*
(Reproduced by permission from BS 5309, Part 4, 1976)

Each riffling stage produces some dust which may cause losses of sample constituents. The amount of dust produced depends on the nature of the material. Care should be taken to include any residual material retained in the slots.

6.4.3 *Method of rotating sample dividing*

The main requirements are that the gross sample should be dry and free falling. Several different types and sizes of rotary samplers are commercially available. This sampling device is suitable for all types of samples including segregates and fine material. The material is fed to the rotating distributor across a vibratory feed assembly (Figure 13). The sample material can be subdivided into two or more portions. The speed of the divider is low at 60–120 r.p.m. and permits a continuous dividing process producing representative portions.[17]

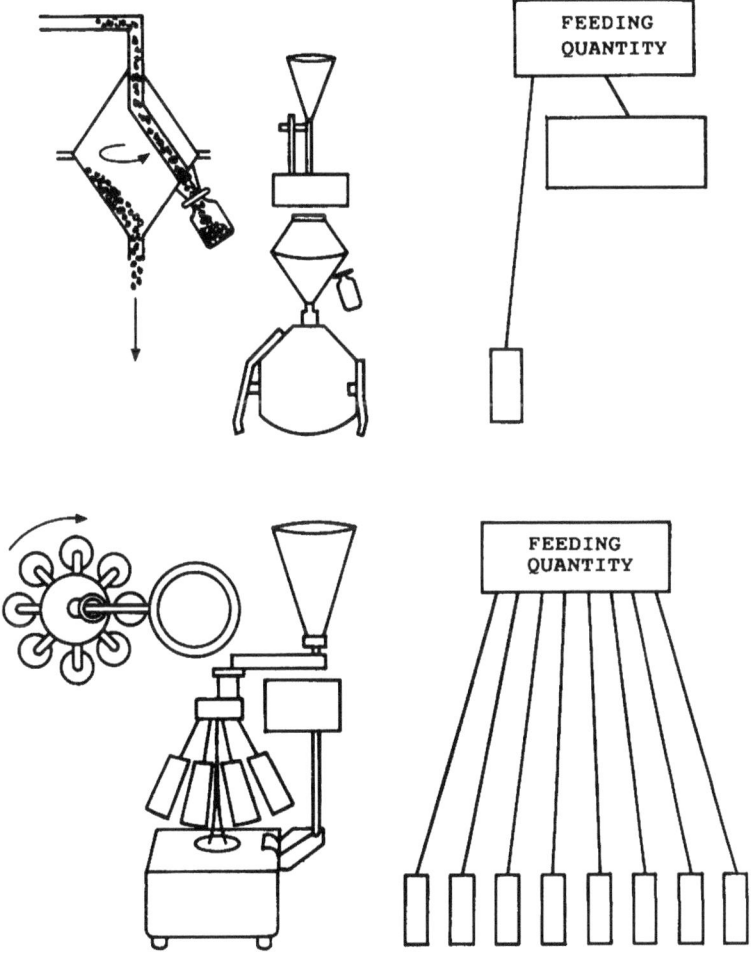

Figure 13

6.5 Investigation of the Reliability of Sampling Methods

Copper in pig feed

Feedingstuffs containing copper, present at a trace level, are fed to pigs. The copper acts as a growth promoter for young animals. Normally, copper is added in granular form as a copper sulfate salt. Even distribution of this salt into a feed material containing several components of various size and density is difficult. The denser salt granules tend to segregate from the bulk material.

Four different sampling methods were investigated for sampling of a feed material mixed with copper salt.

Preparation of the feed sample

A pig feed was ground to pass through a 1.00 mm sieve, and to a 2 kg portion of the feed copper was added, at a level of 250 mg kg^{-1}. To disperse the salt into the feed matrix, the mixture was left overnight on a Turbula mixer (revolving mixer). The feed was then divided into four 500 g portions, and each portion was sampled using the following sampling methods:

1.   Hand Scoop
2.   Riffle chutes
3.   Coning and quartering
4.   Rotating sample divider

Eight increments were taken using each of the above sampling methods, and the metal content was determined by ICP–OES (Inductively Coupled Plasma – Optical Emission Spectroscopy). All the results are illustrated as a bar chart in Figure 14.

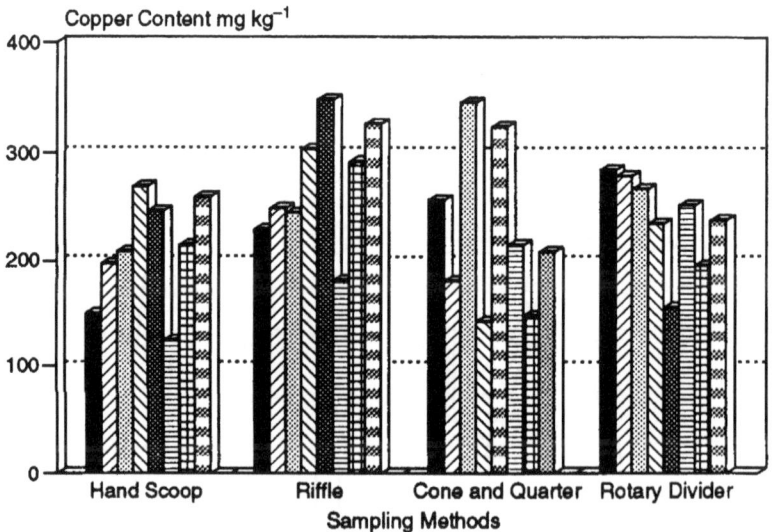

Figure 14 *A comparison of sub-sampling methods: copper content in pig feed*

Table 4 *Copper*

Sampling method	Mean content mg kg^{-1}	Std. Dev.	Coef. var. %
Hand scoop	205.4	48.0	23.4
Riffle chutes	273.5	60.0	20.5
Coning and quartering	232.1	75.4	32.5
Rotating sample divider	244.8	39.6	16.2

The results in Table 4 indicate better results when using the rotating sample divider; the mean value is closer to the expected value and smaller variations are observed between individual results.

6.6 Automated Sampling

In automated sampling a predetermined portion of the material (increment) is taken continuously or at regular intervals. The design of automatic samplers is generally based on the apparatus used for sample reduction but built on a larger scale. Automated samplers are becoming increasingly popular, but they should be rigorously tested before introduction, as they may be subject to bias. Also, after installation they need to be maintained and inspected regularly for bias. Various automatic sampling devices are described in published journals and textbooks.[18]

The primary requirements for an automated sampling device to reduce bias are listed below:

1. The sampling device should take all of the sample, and prevent clogging of the sample material.
2. The sampling cutter should move at a constant rate across the whole of the sample and at right angles to the process stream movement.
3. At the rest position the cutter should be well away from the sample, so that the sample is not withdrawn during the rest period.
4. The cutting interval should be constant if the sampled material varies randomly.
5. The distance between cutting edges should be at least three times the diameter of the largest particle to ensure that largest particles are not excluded.
6. The speed of the cutter should be slow enough (40 cm s^{-1}) to prevent losses due to collision.
7. The depth of the cutter should be at least three times the diameter of the largest particle, to prevent material from escaping.
8. The feed material for sampling should be fed at a constant rate.
9. If possible, the product should be mixed before sampling.
10. The container for sample collection should be sealed during filling to minimise losses of finer particles.
11. The conveyor belt should have brushes or scrapers to remove material on the underside.

12. There should be no contamination or inclusion of material other than the product.

6.6.1 Arc path cutters

These cutters have a chute which is tilted at an angle and mounted on a vertical shaft (Figure 15). The shaft is rotated by an electric motor so that the chute periodically transverses the sample stream.

One disadvantage of arc path cutters is their fixed sub-sampling ratio. Increasing the speed of rotation increases the number of increments taken, but does not give a larger sample.

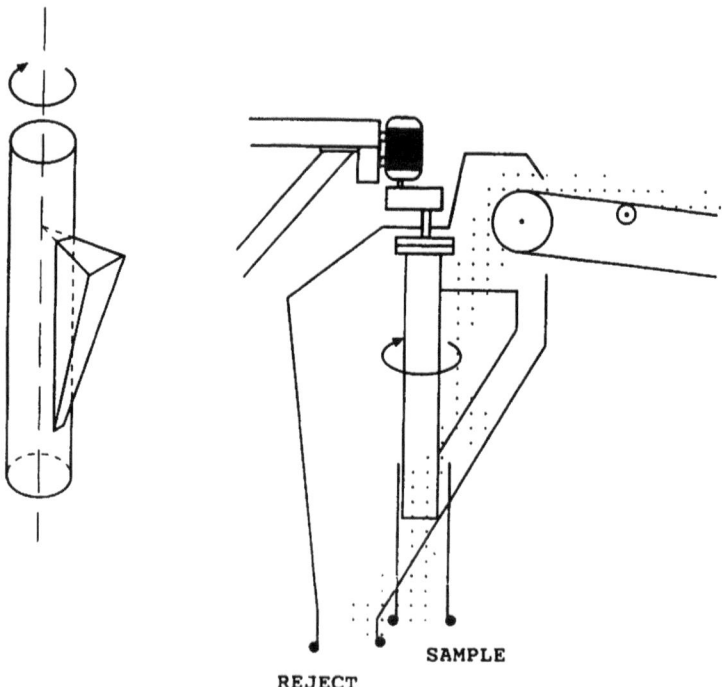

REJECT   SAMPLE

Figure 15 *Working principle of Vezin sampler; inset shows detail of cutter mounted on verticle tubular shaft*
(Reproduced by permission from 'The Sampling of Bulk Materials', by R. Smith and G.V. James, The Royal Society of Chemistry, London, 1981)

6.6.2 Straight path cutters

The straight path cutters consist of a rectangular chute which remains outside the sample stream during the rest position. The edges of the cutter are parallel to the falling stream of the sample material. Straight path cutters allow the rest period to be varied while keeping the same transverse velocity. This allows different sub-sampling ratios to be taken. More increments can be taken by increasing the sample feed rate.

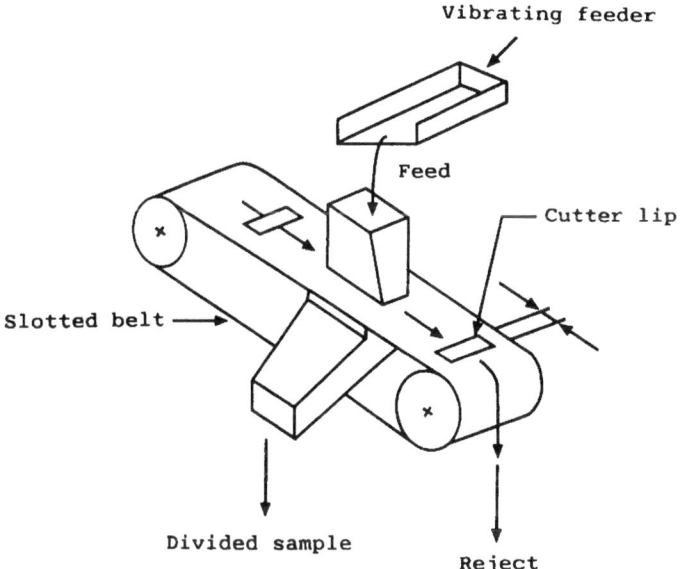

Figure 16 *Example of slotted belt type divider*

6.6.3 Slotted belt samplers

The slotted belt samplers have a conveyor belt with a slot cutter (Figure 16). Material falling though the slot of the conveyor belt is collected in a sample container. The cutter edges should be deep enough to prevent the material overflowing down the chute. However, large particles may bounce on the belt and fall into the sample. Moist samples tend to stick to the underside of the conveyor belt and the cutter edge, hindering brushing or scraping during cleaning.

6.6.4 Moving flap samplers

A deflector moves periodically and cuts the sample across the stream (Figure 17). The flap remains on one side most of the time but is periodically moved to the sampling position for a short interval. This type of sampler is biased in that it always takes more of the stream nearest the sample chute. The bias could be minimized if the flap moves quickly and the sample is collected for a longer period.

Various 'in house' samplers may be appropriate for particular types of sample. Other examples of automatic samplers are given by Block,[19] Sporbeck,[20] and Jordison.[21]

6.6.5 Integrated automatic sampling plant

The primary sample is selected by the integrated sampling plant [22] then reduced by repeatedly milling and screening to obtain a final analytical sample. The

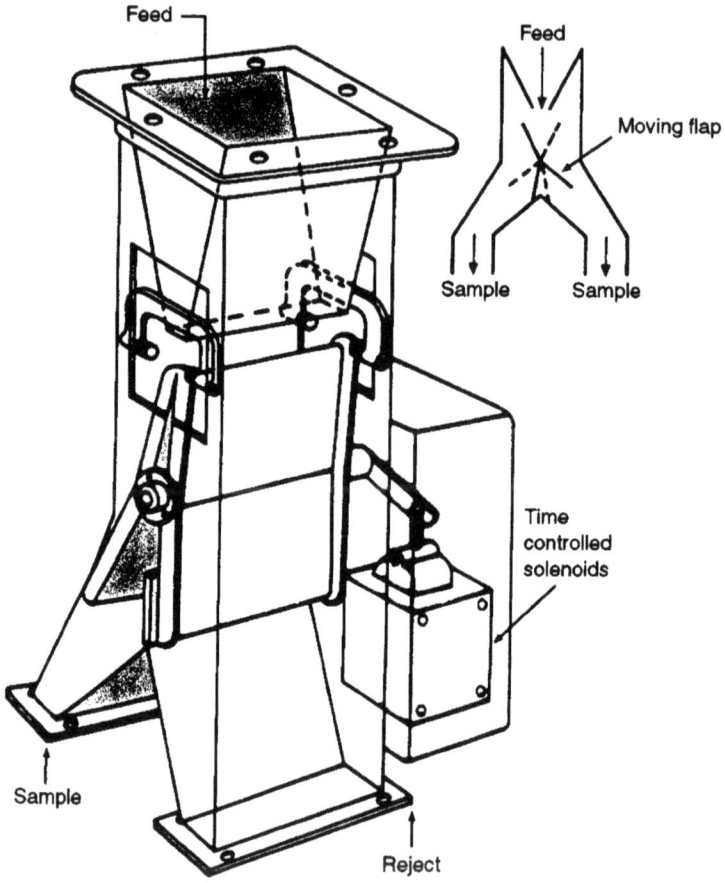

Figure 17 *Moving flap sample divider*
(Reproduced by permission from BS 5309, Part 4, 1976)

advantages of an integrated system are rapidity of operation and convenience. However, fully automated plants are inflexible in dealing with different types of materials and bias in sampling can increase as the plant becomes subject to wear.

6.7 Particle Size Reduction

Particle size reduction can be achieved by applying continuous pressure, by impact, or by grinding with breakers, crushers, or mills. All the equipment used for reducing the particle size of the material should be properly earthed especially when working with combustible materials such as sulfur, starch, and wood which may ignite or explode. In addition, inert atmosphere facilities can also be installed in the grinder to reduce explosion hazards.

6.7.1 Manual methods

Hammer and plate. The sample is crushed with a rounded hammer; this method works equally well with rock, fibrous, and light materials. Material of construction, such as cast steel, generally contains iron and contamination may arise.

Diamond mortar. This is used for breaking very small quantities of very hard material. The shattering fragments from the sample are rotated, and the mortar is made of steel, which reduces the likelihood of contamination. Mortar mills can be used for dry or wet grinding of a wide range of material including diamonds, ceramics, minerals, powders, or paints.

Mortar and pestle. This is the best* method of grinding if contamination is a problem and is convenient for small quantities of product. Mortar and pestle are made from various materials such as glass or porcelain and polished stone.

6.7.2 Mechanical or automated methods

Breakers

Pieces of material larger than 200 mm require breaking before sampling. The following techniques are available:

Hammer: This method can lead to poor sampling if the corners or edges are broken and the interior of the sample is missed.

Drill: This method is useful in that it breaks up the whole lump.

Ball and chain: Here a heavy steel ball is raised above the sample and then allowed to fall freely. This is a laborious and dangerous practice, so appropriate safety precautions are required.

Hydraulic breakers: This is the method normally used for breaking concrete floors in demolition work or for metal ingots.

6.7.3 Crushers

Crushing or size reduction may be achieved by applying a constant pressure, impact or grinding.

Jaw crushers. Jaw crushers can be used intermittently or continuously for crushing samples, which include ore, coal, slag, fertiliser rock, and ceramics, of particle size 20–200 mm. They show little wear with use but the crushing plates may need to be adjusted to suit a particular sampling procedure. Safety precautions (BS 5304[23]) should be taken to guard all the moving parts and the feed aperture should be covered with a flap to contain shattering fragments.

Roll crushers. Roll crushers may have single, double, or four rolls (Figure 18); the single roll crusher works by abrasion. The double or four roll crushers have plate-shaped discs between which the sample material is crushed. These crushers are mainly used for material which is too fine for a jaw crusher and too large for a laboratory mill. The crushing rolls can be adjusted to produce an appropriate particle size of sample.

The maximum size of particles accepted by the roll crusher is derived by Perry.[24]

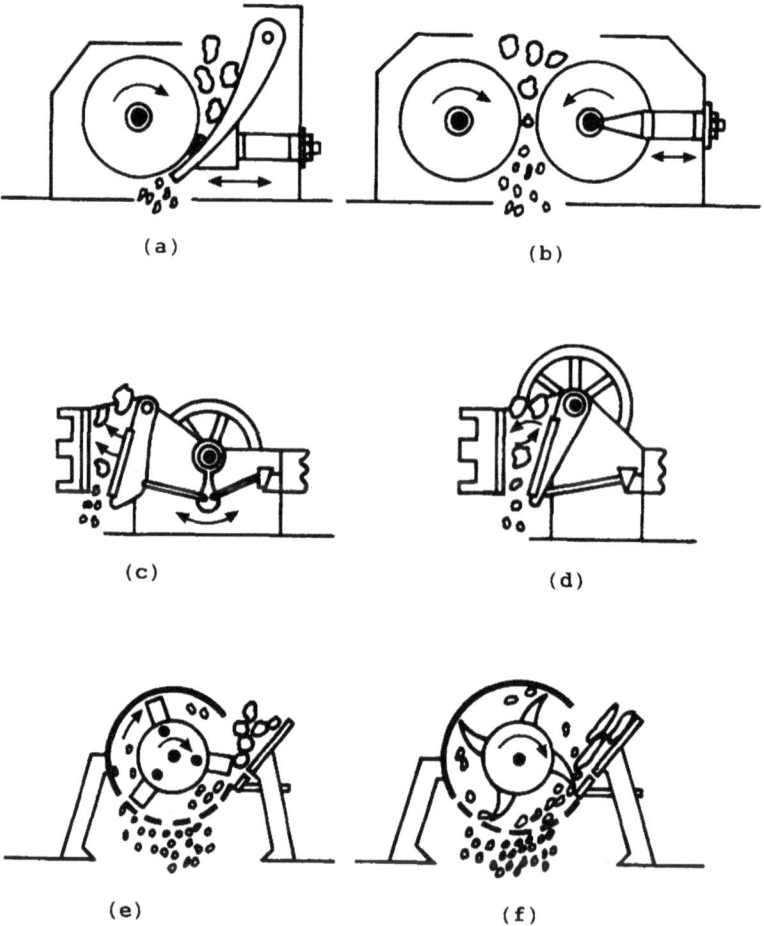

Figure 18 (a) *Single roll crusher with spring-loaded feed plate;* (b) *Double roll crusher with spring-loaded roller;* (c) *Blake pattern jaw crusher;* (d) *Dodge pattern jaw crusher;* (e) *Hammer mill with pivoted flails, adjustable breaker bars, and screen;* (f) *Cutter mill with screen*
(Reproduced by permission from 'The Sampling of Bulk Materials', by R. Smith and G.V. James, The Royal Society of Chemistry, London, 1981)

Cone crushers. Cone crushers have a vertical cone revolving inside a conical crushing chamber wall. These crushers are rarely used in sampling applications on a large scale because of their high cost. Rates of crushing are similar to that of a coffee grinder. The intermediate crushers provide a product of 4 mm or less from a 25 mm feed at rates of several tonnes per hour. The losses of moisture as a result of crushing are generally low but, excessive heat may be evolved with difficult material causing moisture losses and possible decomposition.

6.7.4 Mills

The pan mill consists of one or more rollers revolving in a pan. The pan may be stationary and with moving rollers or *vice versa*. Pan mills are used for material quantities of 50–200 kg with particle sizes of 10 mm or less and reduce the particle size of the material to pass a 1 mm sieve after milling. The mill is easy to clean but some moisture is lost during milling.

The ball mill is a short cylinder filled with balls or rods. The cylinder is made to rotate along its longitudinal axis and the tumbling action of the balls crushes the sample material. The larger ball mills are continuously fed with sample material of 100 mm, which can be reduced to a particle size of 100 mesh (0.1 mm) or less in one pass.

Ball mills are slower than pan mills of similar capacity and can be noisy. Any metallic lead in the sample tends to adhere to the sides and is very difficult to remove. A typical ball mill reduces 50 mm particle size feed to 5 mm and can grind 20–1000 tonnes per hour depending on size.

Ribbon mixers are used for mixing powder such as flour, baking powder, and fairly thin pastes. The mixing element consists of several vertical paddles and two helical ribbons. The material is agitated and moved along to the opposite end of the container. In addition the sample is lifted vertically to allow thorough mixing.

Other types of mills include pug mills for pastes and clays, coffee grinders for cereal grains or seeds, and freezer mills or food blenders and stomachers for biological material such as vegetables or animal tissue samples.

Fine grinding is not possible with some materials (for example food containing fat material), because of their tendency to cake on the sides of the mill. In such cases grinding should not be continued when caking starts because the caked material can alter the overall composition of the sample. The grinding of sticky materials such as gums, resins, and waxes may require a water-jacketed mill or a pulveriser with an air separator-in which cooled air is introduced into the system.

Mills may be designed with screens to allow separation of the finer particles. A cutting mill equipped with sieve inserts can process soft to medium hard substances such as plastic and cellulose based fibrous materials. The grinding chamber has four rotating and three adjustable knives which create a shearing action. Some mills also have a vacuum system where the crushed material is sucked from the mill and blown into an enclosed container.

6.7.5 Closed circuit grinding

The units for closed circuit grinding are extensively used in the metal industry. The devices involve a two stage grinding and consist of a ball mill, a mechanical screen or classifier, and two troughs, for separate collection of coarse and fine particles. The collected coarse material is returned for further grinding.

This form of grinding prevents uneconomical over-grinding and increases the efficiency of the mill by removing the fines which often impede grinding of coarser material.

In some industries even three or four stage sampling is adopted to improve efficiency of grinding.

6.7.6 *Automated equipment for sub–sampling solids*

Machines vary from several tonnes per hour capacity having probe sizes of several cm to laboratory mills giving 30 mesh particles. The larger production scale crushers tend to be noisy, dusty, and prone to wear. The maintenance costs are higher than jaw crushers. Laboratory mills are fast, will handle a wide range of materials, and can be moderately robust. Nearly all machines have either large hold-up volumes or interior contours which do not allow easy cleaning. For this reason, sampling bias may be introduced with certain materials unless the mill is readily dismantled for cleaning.

6.8 Sieves or Screens

A wide range of shakers are available for use with hand sieves. Sieved particles are collected in a bin or plastic sheet. The motor driven shakers impart a circular and tapping motion to the sample.

With all sieves and screens some loss of material occurs which collects in sharp corners and crevices. The loss is unlikely to affect the sample being examined, but it may contaminate a subsequent sample. Sieves should be thoroughly cleaned by brushing, by ultrasonic cleaning, or with a blast of air.

The main uses of sieves in sampling are

(i)   To ensure that all particles have been milled to a particular specification.
(ii) To separate different types of materials and treat them as different sub-samples to avoid problems of segregation.

6.8.1 *Automated sieves*

Grizzlies are used to roll off larger particles in process work and consist of parallel bars held apart by spacer rods. The interval between the rods depends on the size of the fines. The Grizzlies can be flat or inclined and are used before a primary crusher.

Trommels have rotating screens in the form of inclined cylinders for quantities of 100 kg or less. Smaller particles fall through the hole in the screen; larger particles are removed at the open lower end. Several Trommels of different sizes can be placed in sequence for screening wet or dry materials. The slopes of the Trommel vary; greater slope facilitates a thinner bed of material allowing easier passage of fines. Trommels or Reels with wire or fine silk screens are used for sieving finer particles and mainly used in flour mills.

Vibrating screens have rectangular screens for larger quantities of material of more than 100 kg. The larger particles progress to the lower end. The vibrators are adjustable and should be set to give a negligible loss of fines. Multiple units may be stacked on top of one another and are generally used for a range of materials such as coal, soap powder, clay, stone, and various metals. The vibrating screens are highly efficient for coarser particles of larger size than 1 cm, and can also break up lumps, but maintenance is costly. The screens are generally inclined and may be used for both screening and conveying.

Oscillating screens run at low speeds typically, 300–400 r.p.m. and are suitable for finer materials of less than 1 cm in size.

Nylon or Teflon sieves are used for special biological samples to eliminate contamination problems.

7 Sampling of Liquids

Sampling a small quantity of liquid containing one phase is easy because the liquid can be homogenised by shaking. However, in the majority of cases sampling of layered liquids or liquids containing suspended matter is required. Water can collect in a layer below an organic layer, thereby creating a two-phase system. For large volumes of liquids (reservoirs, lakes) the position of sampling is critical. Flowing liquids are also difficult to sample representatively.

7.1 Volume of Sample Collected

The volume of the sample collected is usually determined largely by the number of analytes to be determined and their concentration; larger volumes will be required for trace level measurements.

Sufficient volume of the liquid should be collected for the analyses required and repeat analysis where necessary, with some for future reference.

The following points should be borne in mind:

(i) When contact of the sample with air is to be avoided (determination of dissolved gas, substances that react with air, pH, conductivity), the sample container should be adequately filled but allowing for any expansion due to temperature changes.

(ii) When samples containing, for example, bacteria or undissolved material require vigorous shaking before taking portions for analysis, the sample container should not be completely filled.

(iii) In order to ensure adequate stability of different components, it may be necessary to collect the total volume of sample required in several containers of different types. Amber containers may be required for the measurement of constituents which are sensitive to light. The container or its stopper must not contaminate the sample.

(iv) The containers may require addition of different preserving agents to prevent changes taking place in the constituents of interest between the periods of sampling and analysis.

(v) When the concentration of components such as lead in drinking water changes rapidly with time of flow, the volume of sample collected may affect the concentration in the sample.

(vi) Some measurements such as pH and dissolved oxygen may be necessary before taking the sample.

7.2 Manual Sampling

7.2.1 Sampling of liquids in closed containers (tanks or tank cars, ship cars, drums)

Liquids stored in tanks are liable to stratification due to the differing densities of liquids held within the tank. Liquids in large tanks (road, rail, or ship) should be mixed if at all possible by a mechanical agitator built into the tanks having sufficient area to produce adequate mixing without generating heat which might cause the load to deteriorate.

A variety of mixing devices are available such as turbine or centrifugal liquid mixers. These mixers are generally satisfactory for most liquids, including stratified liquids containing heavy solids suspended in lighter liquids or in highly viscous liquids.

In unmixed liquids the properties are likely to change but, provided sampling intervals are fairly small, systematic sampling should provide good accuracy. However, if at all possible, efficient mixing is preferable to any elaborate system of systematic sampling. The sampling should start from the surface, proceeding to the bottom of the liquid, so that the liquid is not disturbed before taking the sample.

If it is impossible to homogenise prior to sampling, then the increments should be taken at various depths using a bottle in a weighted carrier which can be opened at the required level and sealed after collecting the sample. The cable attached to the sampling bottle is marked to indicate the depth for drawing the sample. It is essential to ensure that the final sample contains the sub-samples combined in proportion to the volume it represents. This procedure is appropriate when investigating abnormalities or when the contaminants have the tendency to concentrate at the top or bottom layer of the liquid. In certain cases it may be necessary to analyse the sub-samples separately.

A procedure which is suitable for small containers is to sample the complete depth of the liquid. An open tube sampler is placed into the container, allowed to draw up the liquid, and then closed (Figure 19).

An alternative device which is more appropriate for inhomogeneous liquids is illustrated in Figure 20. The device is lowered to the required depth and the liquid is drawn into it by raising the plunger.

For tanks, a special sampler similar to a thief sampler can be used. This device has a valve which is opened when the sampler touches the bottom of the liquid and is closed when the sampler is raised.

Liquids in tank cars or ship cars can be sampled while unloading, a side arm outlet from the main unloading pipe is automatically opened at intervals and a small amount of liquid is collected and mixed.

When the quantity of liquid to be sampled is in more than one container, a

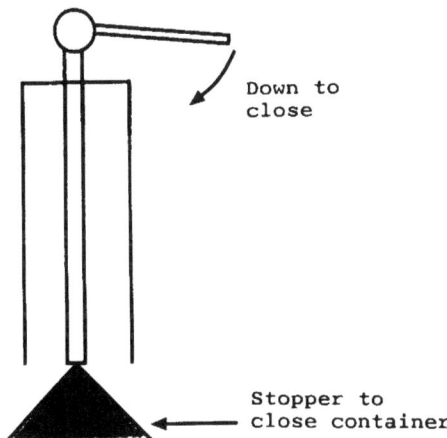

Down to
close

Stopper to
close container

Figure 19

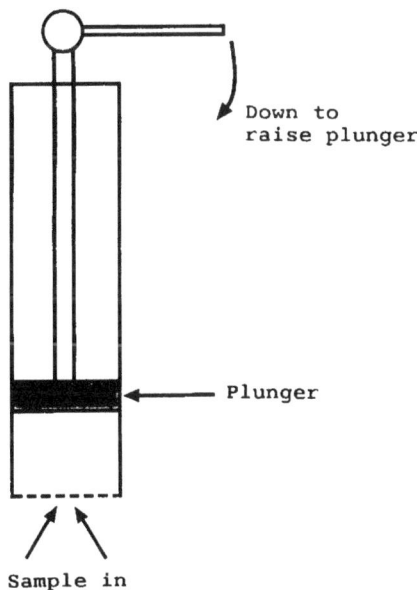

Down to
raise plunger

Plunger

Sample in

Figure 20

representative quantity of liquid is taken from the selected containers, then mixed to form the bulk sample. All the containers should be examined for any obvious differences such as batch numbers, colour, *etc.*

7.2.2 Sampling of liquids in open systems (rivers, streams, canals, industrial effluents)

The chemical composition of a flowing liquid may vary according to changes in a number of parameters such as temperature, flow rate, distance from sources, none of which can be controlled by the sampler. Because of these factors it is difficult to draw accurate conclusions from a single sample. The full information may be available only after a large number of samples have been taken and analysed.

The samples are commonly collected by using wide necked bottles or canisters which are immersed at a suitable point in the flowing stream. For sampling at various depths a weighted glass bottle with a removable stopper is a simple but effective device.

Dipper sampling can be applied when there is a free flow of the liquid in the stream. The full cross-section of the stream is collected at fixed intervals with the dipper.

The sampling device used for bacteriological samples should facilitate sterilisation and be constructed from an inert material. It may be advantageous to use a sterile sample bottle in the sampling device when sampling at depth for the bacteriological examination of water.[25]

Sampling systems for natural waters such as rivers must be carefully selected and installed to avoid blockage due to debris in the water. The inlet can be protected with both a coarse and a fine mesh and the debris collected should be removed frequently. When pumps are required to deliver a stream of sample, water level fluctuations in rivers and estuaries may cause installation problems.[26] Mounting the pump on a floating platform has often been used to overcome this problem. A detailed report on sampling for the determination of water quality has been published.[27]

It is very important that the procedure to be used for collecting the sample is carefully described in the sampling plan. Contamination of the sample from the environment around the outlet of the sampling line should be avoided especially when the components of interest are present at a trace level.

Care should be taken in choosing the sampling locations: sampling at or near the surface, bottom, banks, and stagnant areas should be avoided. It has been suggested[28] that samples from a stream, whenever possible, should be collected 30 cm below the surface and at a similar distance above the bottom.

It is a useful practice to rinse the sample container with the sample two or three times before collecting the sample. However, this practice cannot be employed when the sample needs to be collected with a preserving reagent or when the sample contains materials such as suspended solids, metals, oils, and grease that may be adsorbed on the walls of the container. In such cases, the container must be dry and clean.

During winter the lakes may be covered with ice. Therefore the collection of samples for dissolved oxygen content requires particular care to prevent contamination of samples by air. Sample bottles should also be adequately filled, allowing for expansion, and securely sealed so that oxygen is not lost from the sample as the temperature rises.

7.2.3 Liquids flowing within closed systems (filling lines, pipelines)

The flow rate of the liquid is quite important. At low flow rates laminar flow predominates; as a result the liquid in the centre of the pipe flows at maximum velocity and the velocity decreases to zero at the wall. To ensure homogeneity, a turbulent flow can be created just before the sample is taken. To avoid the sample being forced out of the sampling tube by the flow of the liquid, the sample should be removed from the direction opposite to that of the flow, if at all possible. If the liquid to be sampled is semi-liquid, the pipelines and receivers may be warmed to keep the material just above the liquefying temperature.

7.2.4 Liquids in open locations (lakes, reservoirs)

Weighed glass bottles can be used to sample large volumes of liquid in static situations. The bottle (grab sampler) is lowered into the water, *e.g.* a lake, to the required depth, as measured by the length of the rope or chain holding the basket. The stopper is removed and the sample taken. The device is then hauled back to the surface and the sample is transferred to a suitable container for storage. Many such devices are used and a number of these are described.[29,30] It is important to select a sampling device which is robust so as to withstand rough handling and if used at great depth to withstand high pressures.

If the lake or reservoir is shallow, and the sampling is to be carried out on a regular basis, then a permanent sampling device is useful. The sampling device should be constructed from inert plastic and allow increments to be taken at various depths, to produce separate or composite samples.

The sampling position is easy to choose when the quality of water flowing out of a lake or taken from a reservoir is to be measured. The stream water flowing into a lake or reservoir from various sources may differ to a greater degree and it is often useful to sample these streams individually to determine the quality of water.

The main problem in selecting sampling locations arises when it is necessary to determine the quality of water in particular areas of a lake or reservoir or throughout the whole water system. In such cases of heterogeneity quite detailed investigations from several locations will be required.[28]

During the summer, as the surface layers of a lake become warmer, the density of the water decreases, and vertical stratification tends to occur which progressively reduces the concentration of dissolved oxygen in the lower layers. The bottom layer may eventually become anaerobic, releasing other substances such as ammonia, iron, manganese, nitrate, and phosphate from the bottom sediment. In addition growth of algae in the surface layers of water may cause changes in pH, alkalinity, hardness, and in the concentrations of ammonia, nitrate, nitrite, phosphate, carbon dioxide, and oxygen.

7.3 Automatic Sampling

Automatic equipment reduces the manpower required, eliminates the need to expose personnel to hazardous conditions, and allows the collection of samples outside normal working hours. These samplers are increasingly used for many applications such as the preparation of composite samples, monitoring quality, and obtaining samples at inaccessible locations. Some samplers are portable and can be invaluable for many purposes.

The sampling devices can be set to collect a series of samples at specified intervals, into separate receivers, or into a single receiver. Two review papers[31,32] include more details. When considering the use of automatic samplers, their suitability should be critically assessed. Parameters such as the flow rates within the sampling lines should be optimised for a particular application.

7.4 Filtration of Liquids

It may be necessary to filter the sample where either the soluble or insoluble components are to be measured. In some cases the suspended matter may interfere or cause changes in the constituents being determined and therefore needs to be removed. All the equipment used for filtration should be carefully checked to ensure that no contamination is taking place and that losses of the constituents of interest are not occurring through absorption or adsorption on filter papers or the equipment being used.

7.5 Preservation of Samples

Stability of many constituents can be achieved by addition of a chemical reagent to the empty container before collection of the sample or immediately after collection. The preserving agent may interfere with the analytical method; thus initial tests must be performed to check the compatibility of the preserving reagents. Acidification improves the stability of water samples requiring trace metal analysis, but the minimum acidity and the type of acid required depends on the metal to be determined.

Polynuclear aromatic hydrocarbons are found at very low levels, generally in the order of ng l^{-1}. At such low concentrations, adsorption onto the surface of the container causes a major problem. Addition of the extracting solution prior to the transference of the sample to the container improves the recoveries of these compounds. Biocidal reagents may be added when the constituents of interest, such as nitrogen and phosphorus compounds, are subject to biological reactions. Special care is required for sampling to determine phosphorus, as orthophosphates and other constituents are released when biological species are killed. Many papers have been published on comparisons of preservatives.[33-35]

Some volatile compounds are conveniently stored with the aid of a chemical reaction to form a less volatile product, *e.g.* hydrogen sulfide can be converted into zinc or cadmium sulfide. When no suitable reaction exists, the compounds may be adsorbed onto an inert substance such as activated charcoal and recovered when required for analysis.

8 Sampling of Gases

Gases can alter in their overall composition due to changes occurring in their temperature and pressure. Thus the storage of gases may present difficulties: when a sample is taken at a pressure or temperature that is higher than ambient, condensation of less volatile components (*e.g.* water) can occur. Therefore, in order that the sample may return to its original state, it is necessary to heat the sample until equilibrium is achieved. Gases which are at high pressure when they are sampled are best stored at the same pressure to avoid condensation. In order to avoid the storage problems, gases are often directly transferred to the analyser to measure the composition or properties of the gas. Factors such as unexpected atmospheric contamination or chemical reaction and high humidity may reduce the sampling efficiency. It is important to test the validity of any sampling method on site to determine the accuracy of the method.

The monitoring of hazardous gases in the atmosphere may be undertaken to assess the exposure of employees to the pollutants which may contain components in gas, vapour, particulate, and/or aerosol form. General classification of the pollutants may vary depending on their definition in various literature sources.

Gases: These are generally classified as molecular in size, about 0.002 μm in diameter, and do not condense at room temperature.

Vapour: This is produced from volatile liquids and can start condensing if the concentration is high.

Particulate matter: Any material which exists as a solid or liquid in the atmosphere and occurs in a range of particle sizes is classed as particulate matter. It may consist of solids (coal, wood) which are classed as dust, or liquid droplets (oil, water, solvent) which are classed as mist.

Aerosols: Aerosols are a group of particles in either solid or liquid state which are small enough to remain suspended in the atmosphere.

8.1 Sampling Plan

Generally the gases are delivered by pipeline apart from compressed or liquefied gases which are delivered in cylinders. Thus the sampling is related to the flow rather than a specified volume from a lot or batch.

The purpose of sampling will in most cases define the type and the volume of the sample required. The sampler may be required to provide a spot, a composite, an intermittent, or a continuous sample. The procedure adopted for the monitoring of air pollutants in terms of duration, location, and frequency of sampling will depend on various factors such as the objective, the nature of pollutant, sampling device, and the measurement system to be used.

8.2 Liquefied Gases in Cylinders

The liquefied gases may be sampled either as gases or as liquids using a variety of pressure vessels. During evaporation losses of more volatile gases occur while the less volatile constituents condense. Reducing the pressure of liquefied gases will cause evaporation of volatile constituents. Samples collected in a warm receiver such as a Dewar flask lose volatile constituents. Therefore it is good practice to flush the container and sampling lines before taking the sample. The content of each cylinder is generally uniform, so sampling provides an indication of the variation between containers.

8.3 Gases in Storage Tanks

8.3.1 Static gas (stratified)

Gases stored in large tanks often separate into layers (stratified) owing to differences in density. Thus mixing or equilibrium by diffusion may be required to produce a sample representative of the entire contents. Normally spot samples at random or continuous intervals are taken to give an indication of the extent of stratification. Impellers or circulation fans may be used to provide mixing before sampling.

8.3.2 Gases in motion

Gases from a gas mains or from chemical plants have high (turbulent) flow rates and due to natural diffusion can be considered homogeneous, with respect to the cross-section of the unit. The homogeneity of the sample can be checked at specified intervals with a sampling probe which is directly linked to an analyser or a collecting vessel.

8.4 Sampling from the Atmosphere

8.4.1 Grab sample

An air sample is collected in a flask, bottle, bag, or any other container. A single sample may be collected to assess the quality of air, or several sample collection devices can be placed at different locations in a large area.

8.4.2 Continuous sampling

A sample is collected by adsorption or absorption onto a solid (silica gel, charcoal) or liquid medium over a period of time. The sampling can be carried out with the use of a pump by forcing a flow of air through a sample collector or by letting the air diffuse through the medium. A wide range of samplers are available; the simplest are gas wash bottles such as a Dreschel, an impinger, or a bubbler.

8.5 Sampling Equipment

The basic equipment consists of a sampling line, a probe or detector, collection vessel (flask, bottle, bag, impinger, bubbler, adsorbent tube), and an aspirator and other accessories such as filters, pressure regulators, or safety valves.

8.5.1 Probe or detector

The probe or detector should be robust and able to withstand the temperature and pressure of sampling. It should not react with any of the constituents in the sample and should be chemically inert.

8.5.2 Sampling lines

Sampling lines should be short and chemically inert and impermeable to the gas being sampled. A filter assembly may be included to protect the lines from dust, moisture, or other constituents which may cause blockages.

8.5.3 Sample containers

The sample collection vessels vary in size, and are constructed from glass or a suitable material which is inert towards the components in the gas sample. The vessel is evacuated so that when the seal is opened gas enters as a result of vacuum. Alternatively, a pump may be used to draw the sample into the container. Gases containing acidic components should not be stored in metal containers as losses can occur due to chemical reactions with the metal surface. If the gas sample is at all wet, then corrosion of the container is likely. Metal containers tend to be less inert than those of glass or plastic material. Gases at high pressure require suitably constructed pressure vessels.

Samples should be analysed as soon as possible after collection. If any delay is likely to occur then prior storage tests should be carried out to ensure that the sample is analysed before decomposition of the sample constituents takes place.

8.5.4 Absorption by impingers

Atmospheric gases or vapours can be collected by absorption into a liquid solution contained in an impinger. Absorption can be improved if the absorbing compounds are reactive and form a stable non-volatile compound with the sample. The impinger consists of a graduated receiver and a tapered inlet tube, the tip of which should be in contact with the liquid in the receiver. If the extraction efficiency is less than 100% due to incomplete absorption then appropriate corrections should be made.

Portable impingers are also available. These are often used to monitor the breathing zone of personnel in workplaces. The impinger is attached to the collar of the employee but should be held in a vertical position to prevent spillages of liquid. A trap should be located in the impinger to collect the liquid if any accidental spillage does occur. Some non-spill impingers are also available which may be more suitable for certain applications.

Impingers can also be used for the sampling of particulates in polluted air. These can be placed in series to ensure that all of the sample is collected. In certain cases a dry filter may precede the wet impinger; for example, when collecting sulfuric acid fumes, the finer particulate matter may escape from the impinger but can be collected by the filter.[36]

A midget impinger containing 1-(2-methoxyphenyl)piperazine absorbing solution has been used to determine organic isocyanate in air.[37] Aldehyde and ketones in atmospheric gases may be determined by absorbing into 2,4-dinitrophenylhydrazine derivatives. Published methods are available for the determination of arsine[38] and a field method for the determination of formaldehyde[39] in air.

8.5.5 Absorption by bubblers

The sampling of gases and vapours which are less soluble in the collection liquid is improved if sampling is carried out using a fritted or sintered glass bubbler. The sampling is performed with a bubbler such as a Dreschel bottle. The air is drawn in through an inlet tube forming bubbles; this allows better contact between the air and the liquid reagent in the bubbler. Absorption is efficient if the size of the bubbles is small; this can be achieved by the use of an inlet tube with a small diameter orifice.

The concentration of the pollutant can be determined, by a simple equation, from the volume of air sampled and the mass of the pollutant collected on the absorber.[40] Measurement is also possible if two or more absorbers are used in series to collect different pollutants and the air is sampled sequentially through each absorber.

Portable bubblers are often used to monitor the air quality near the breathing zone during a shift. The sampler is located at the collar of the subject or near the subject's breathing zone. Sampling can be undertaken over a 24 h period with observation at several times during this interval.

Various combinations of bubblers and chemicals may be used for the measurement of toxic compounds in air.[41] Collection of pollutants by bubblers has been used for measuring a wide variety of inorganic and organic components in locating the source of contamination in industrial and atmospheric samples. Sulfur dioxide has been collected by absorption in hydrogen peroxide.[42,43] A reference method used in the USA and EU countries utilises a mixture of tetra-chloromercurate/pararosaniline for the absorption of sulfur dioxide. Other absorption methods exist for the collection of most atmospheric pollutants, such as hydrogen sulfide,[44] chlorine,[45] mercury vapour,[46] and ozone.[47]

8.5.6 Filters

Pollutants from the atmosphere can be passed through a filter material, which concentrates the air sample. Collection with filters offers several advantages in that the system is more robust and easier to transport, and several pollutants can be collected sequentially with different types of filters. To sample the air a filter is placed in a holder and connected to a sampling pump to draw off the air for a fixed period.

In sampling for atmospheric pollutants the presence of aerosol particles may cause interference; therefore it may be necessary to remove the aerosols by a prefilter. However, losses of reactive gases by adsorption onto filter material can occur. Glass fibre filters can be used to collect sulfur dioxide and nitric acid with high efficiency. Nylon filters are highly efficient and react with 90% of the nitric acid and hydrochloric acid.[48-50]

In order to collect specific components of air, filters impregnated with a chemical reagent are preferred. The chemical coated on the impregnated filter reacts with the sample to form a stable compound. The variability in the efficiency of impregnated filters is frequently a problem when sampling atmospheric contaminants. Sampling efficiency of the filter is dependent on the gas velocity across the filter, the concentration of the sample components, and the humidity of the atmosphere.[51] Sampling should not be continued after the exhaustion of the filter.

8.5.7 Spot test papers

Monitoring devices which rely on spot chemical or biological reactions can be used to monitor toxic gases. Test papers coated with chemical reagents which change colour in the presence of toxic gases may be used.[52] The intensity of the colour is related to the concentration of components. The gas is drawn through the paper with a hand pump or any other automatic sampling device. The colour measurement is performed with an optical reader to give a direct concentration value or compared against a standard colour chart. Test papers are available for the measurement of a wide range of organic or inorganic compounds including, chlorine, phosgene, vinyl chloride, sulfur dioxide, ammonia, nitrogen dioxide, and nitrate.

The presence of hydrogen sulfide gas can be detected by the discoloration of paint or with lead acetate paper. Some commercial systems use this to provide concentration alarms for hydrogen sulfide. Actual concentrations cannot be measured with this method, but an overall indication can be obtained. Lead dioxide candles (now obsolete) were formerly used to monitor sulfur dioxide: the coating changed colour on formation of lead sulfate.

8.5.8 Spot detector tubes

Indicator tubes provide a rapid, cheap, and simple procedure to test the presence of toxic gases and vapours. The system consists of a long, narrow cylindrical detector tube, through which the air sample is drawn via a pump. The internal walls of the tube are coated with a chemical reagent which reacts with the gaseous pollutants to form a coloured stain. Alternatively, the detector tube may be packed with an inert support material impregnated with the chemical.

Monitoring of toxic gases in workplaces is often performed with indicator tubes which contain a support material such as silica gel coated with a reactive chemical. The reaction between the gas and the reagent forms a coloured product, and the concentration of the the gas is indicated by the length of the stain or the intensity of the colour. Detection limits for most of the gases tend to be above 1 p.p.m.; however, some non-specific reactions may take place if organic compounds of similar chemical properties are present.

Detector tubes are suitable for measuring a wide range of organic (alcohol, ester, hydrocarbon) and inorganic (nitrogen dioxide) compounds. Indicator tubes should only be used for monitoring of the compounds specified by the manufacturer. Tubes with a secondary coating are also available which remove interfering components, these are more appropriate for the sampling of a mixture of gases.

This technique has been used to collect several gaseous atmospheric contaminants.[53,54] Sulfur dioxide has been collected by a detector impregnated with lead dioxide, potassium carbonate,[55] or tetrachloromercurate. Ammonia has been collected in tubes coated with oxalic or phosphoric acid.[56]

8.5.9 Adsorption by columns

Adsorption is a surface phenomenon which involves concentration and bonding of sample molecules onto a solid surface. Sampling by adsorption onto a solid inert support and separation by gas chromatographic techniques can be used for the analysis of atmospheric gases.

When selecting a suitable adsorbent polar interactions must be considered: non-polar sorbents will adsorb non-polar organic phases but exclude polar compounds such as water vapour. In addition the adsorbent chosen should collect all the components of interest within the breakthrough volume without losses. The maximum volume of air that can be sampled without any losses is governed by various factors including humidity, volume of air, and the concentration of components in the atmosphere.

A wide range of solid materials with extremely porous surfaces has been employed to collect the pollutants. Activated charcoal, silica gel, and organic polymers are the most common adsorbents and have been used to sample a variety of organic compounds in the atmosphere.[57,58]

Activated charcoal is often used to collect vapour samples, principally for industrial hygiene purposes,[59-61] and the literature includes several official methods published by the Health and Safety Executive (HSE)[62] and the National Institute for Occupational Safety and Health (NIOSH).[63]

Silica gel is less reactive than charcoal and more suitable for the sampling of polar organic compounds. However, the presence of excess moisture in the atmosphere can decrease the sampling efficiency. The adsorption capacity has been improved by impregnating silica gel with 2,4-dinitrophenylhydrazine to collect acetaldehyde and formaldehyde; the mixture is extracted with acetonitrile and analysed by high pressure liquid chromatography.[64]

A polymer adsorbent Tenax GC [poly(2,6-diphenylphenoxy ether)] may be used for the sampling of a variety of organic compounds at trace levels. Tenax sorbent tubes can be used for the sampling of glycol ethers and glycol acetate vapours in air.[65] Tenax GC has been used for sampling benzene in air,[66] chlorinated hydrocarbons, and a variety of other volatile and sulfur compounds.[67,68] This polymer has also been used for the collection of a mixture of 27 organic compounds; the recoveries were above 85%.[69]

The choice of the adsorption system depends upon the characteristics of the pollutant. If a range of compounds with different polarities is of interest, then sorbent tubes may be connected in series to collect all the components.

To collect the atmospheric sample, break the seal of the tube, place it in a vertical position in its holder, and draw a measured volume of air through the tube. Cover with a plastic cap to dispatch to the laboratory for analysis.

Subsequent analysis of the adsorbed pollutants requires either a solvent extraction or thermal volatilisation at high temperature. For thermal desorption the collected vapour is desorbed by heat and is transferred under inert carrier gas into a gas chromatographic system with detection by flame ionisation. This procedure is often more sensitive and less time consuming than the solvent extraction system.

8.5.10 Diffusion sampling

Diffusion sampling is a cheap and convenient method for sampling a wide range of gases and vapours, and can be used to take samples from inaccessible places. A bottle or a container fitted with a screw cap is suitable for taking samples. To collect a sample in the container, the screw cap is removed and the container is placed in the sampling location for a fixed period (*e.g.* 30 min) and then sealed.

Diffusive samplers containing a membrane or adsorbents such as charcoal can also be used for sampling gases and vapours.[70-72] The sampler is placed in contact with air for a fixed period to allow the components to adsorb onto the solid surface. These types of sampler are available in badge or tube form for monitoring personnel in the workplace. Detailed instructions to check the performance of the diffusive samplers are given in reference 73.

8.5.11 Portable instruments

A wide range of instruments is available for on-line sampling and/or analysis of gases. Electrochemical sensors are employed to monitor the levels of oxygen, sulfur dioxide, hydrogen sulfide, carbon monoxide, and toxic gases in the air. These sensors are essential when the presence of hazardous gases is suspected. The detectors consist of an electrolytic cell, into which the air sample diffuses through a membrane. The sensors can give a digital readout of the concentration, and some are also equipped with alarms to give an indication of hazardous conditions. The alarm should give both audible and visual warning. Total combustible gas indicators are employed for the detection of organic solvent vapours.

Various spectroscopic instruments based on colorimetric, ultra-violet, and infra-red detection may be used for monitoring gases[74] (flammable gases methane, ethane, butane, *etc.*). The air sample from the location of interest can be pumped directly into the sample cell of the detector for the continuous monitoring of the atmosphere. Alternatively, the detector can be set to give a reading at fixed intervals.

Portable gas and high pressure liquid chromatographic instruments are also available for the screening of gases. Further analysis in the laboratory is usually carried out to confirm the findings.

9 Containers

9.1 Choice of Sample Containers

It is extremely important to ensure that the test sample has not been contaminated or transformed in any way during sampling, transport, or storage. The sample container may have an important effect on sample stability and analytical methods often make special recommendations on the type of container suitable for each component. Sample containers should be clean and constructed of material that is inert to the substance being sampled.

Metal, glass, or plastic containers (PTFE, polypropylene, *etc.*), fitted with airtight closures which are capable of preserving the sample, are commonly used. Glass bottles are transparent and have the advantage that the condition of the sample is more readily apparent; they can also be thoroughly cleaned and sterilised for bacteriological samples. It is important to choose cleaning solutions or detergents which do not contain the constituents of interest to minimise contamination. Opaque sample bottles may be used where photochemical reactions need to be reduced. In certain cases, plastic bottles may be preferred when the samples need to be frozen for storage and transport. Plastic is less prone to breakage than glass.

9.2 Contamination from Sample Containers

To avoid excessive sample handling, one appropriate container should be used for each sample, preferably one in which the sample is placed immediately at the time of collection. The sample should be stored in its container and isolated from the environment, following transport from the sampling site to the analytical laboratory, prior to analysis and long term storage. Freezing and/or maintaining the sample at temperatures as low as feasible will in most cases minimise concentration changes. However, the possibility of components crystallising out at low temperatures or on freezing should be considered.

When polythene bottles are used volatile substances and oxygen diffuse slowly through the walls; thus contamination by the gases from the air around the bottle may occur. The materials from which the sample container is made may cause contamination of the sample. Sodium, silica, and boron can leach from borosilicate glass, and organic components such as plasticisers from plastics. Certain constituents of the sample may react with the container material; for example, fluoride may react with glass. These effects become more important when the constituents of interest are present at trace level.

Stoppers and caps should be made from the same material as the sample container. To minimise contamination the use of rubber or plastic inserts should be avoided where possible. For containers requiring a tight seal plastic inserts may be necessary, but these should be tested prior to use in order to avoid contamination.

The nature and the extent of contamination may depend on the manufacture of a particular type of container. Furthermore containers of identical type from the same supplier may also differ.

9.3 Adsorption on Sample Containers

Other factors to be taken into account when choosing the sample container are that the sample components may be adsorbed onto the walls of containers: trace metals adsorb onto glass surfaces and organic compounds such as benzene adsorb onto plastic materials.

For constituents such as various types of oils, greases, and organochlorine pesticides, it is virtually impossible to prevent adsorption onto the surfaces of the sample container. Therefore, it is necessary to reserve the sample container for individual constituents, so that the adsorbed material can be removed from the container as part of the analytical procedure.

9.4 Sample Containers for Biological Samples

Changes in the form and concentration of the numerous trace constituents in the sample can occur in several ways. Processes such as adsorption, biodegradation, and permeation may reduce the concentration of various components. Continued biological activity can produce species that may not have been present in the original sample or destroy the components of interest.

Sample containers must consist of inert material; for biological samples Teflon appears to be most suitable if both organic and inorganic traces are being considered. Trace constituents in Teflon are extremely low; thus leaching of the contaminants is minimal. In addition, the absorption and diffusion rates for Teflon are low, minimising the loss of trace constituents from the sample. PTFE containers are less costly and may be used in certain cases.

A blank container from the same batch must be checked for the constituents to be determined and all the sampling implements must be evaluated periodically.

10 Information to be Submitted with the Sample

10.1 Documentation

Information on the conditions during sampling, transport, storage, and preparation for analysis must be collected and stated. Any unusual occurrence should be documented to enable the analyst to evaluate any unusual results. Unambiguous instructions should be prepared for all the personnel involved in the sampling programme to ensure reproducible sampling.

A number of other details such as sampling locations, methods of sampling, preservatives added, analysis required, name of the sampler, and date and time when the sample was taken should all be recorded. The report for the sampling of foods[2] may also make reference to the condition of the product sampled, including signs of insect, mite, rodent, or other infestation visible in the warehouse, silo, or mill during work on the vessel or carrier. This infestation is not always readily apparent in the sample except on close inspection.

10.2 Labelling Sample Containers

Sample containers and lids must be clearly labelled so that subsequent analytical results can be properly interpreted. All the details relevant to the sample should be recorded on the label attached to the sample container. If a large batch of samples is collected, then it may be easier to identify the containers by a code number or a bar code and to record other details on a separate form. All the labels or forms should be completed at the time of sample collection.

11 Transport and Storage

A vital consideration in sample transport and storage is in ensuring that the concentration of constituents of interest in the sample does not vary significantly or in an unknown way between the time of sample collection and analysis. Some samples may be prone to degradation and in many cases it may be necessary to analyse the sample as soon as it is collected. However, immediate analysis may be impossible so appropriate procedures need to be adopted for transport and storage.

Many chemical, physical, and biological processes as well as the conditions under which the sample is handled, such as temperature, exposure to light, freezing, the composition, the dimension of the sample container, and agitation during transport can lead to marked changes in the composition.

11.1 Transportation

Collected samples should be transported as soon as possible to the laboratory for analysis. Facilities for the refrigeration or even freezing of samples during transportation may be essential for unstable constituents. To prevent compositional changes samples should not be transported at temperatures above which they were collected. When samples cannot be brought to the laboratory sufficiently rapidly, the use of on-site analysers for particularly unstable constituents should be employed.

Natural solid samples (ores, rocks) are very stable and can be stored and transported in containers made from various materials. However, care must be taken to ensure that fine particles which may be broken off during transport are retrieved. Thus, these samples are often wrapped in plastic-coated paper or plastic sheet before packing into wooden crates.

The transport of flammable liquids should be carried out using mobile fire resistant storage cabinets. Appropriate precautions should also be taken when transporting samples consisting of radioactive constituents.

11.2 Special Storage Conditions

11.2.1 Refrigeration

The storage of certain samples in the dark at low temperature (normally 4 °C) is a commonly used procedure for sample preservation. At low temperature, the rate of chemical reaction and biological activity is significantly reduced which minimises the decomposition of sample components. It is desirable to refrigerate the sample immediately after collection especially if the constituents of interest are unstable. Changes occurring in sediments and sludges can be reduced by refrigeration.

11.2.2 Freezing

Freezing of samples (-20°C) in polythene bottles immediately on collection is desirable for biological (clinical, food) and water samples. If this is not possible then the analysis of the sample must commence before any changes in the composition occur. Repeated thawing and freezing should be avoided as these may result in losses of some constituents. Freezing has been recommended for the storage of water samples for periods of up to several weeks for the determination of nitrogen, phosphorus, and silicon compounds.[75]

Shock freezing is useful for particularly unstable biological samples immediately after removal from the natural environment. The mobility of the constituents and chemical reaction rates are greatly reduced at lower temperatures of -120 and -196°C. Vapour phase liquid nitrogen freezers appear to be most suitable, since no mechanical or electrical equipment is involved in the cooling process. The nitrogen vapour provides a controlled, chemically inert environment for the samples. The use of nitrogen after collection and during transport and storage minimises changes in temperature which may lead to segregation of components. Liquid nitrogen vapour phase transport containers are accepted by all commercial carriers to allow rapid transport of samples from the sampling site to the analytical laboratory.

Freezing must be avoided with certain samples such as emulsions, where the recovery of a representative sample is impossible.

Appendix 1: Theoretical Aspects

Statistical data on sampling furnishes an estimate of the precision which may apply to accidental variation such as variability between units, variability of the test from day to day and from hour to hour, and small independent random errors in analysis. Quality and reliability of data are built in through a proper design of the sampling programme, taking into consideration the theory of probability.[76]

Most investigations of a sample, whether they are concerned with quality control or with chemical composition, have to take into consideration two points in order to draw correct conclusions from the experimental results.

(i) The quantity of the material actually examined is a tiny fraction of the whole sample for which the information is required. The sampling scheme must therefore attempt to allow for the possible variations within the sample.

(ii) For most constituents of a sample, the procedures for collection, analysis, and examination of the samples are not free from error. The accuracy required may be limited by the technology or the cost.

The sampling protocol may include the following information:

(i) The statistical criteria to be used for acceptance or rejection of the lot on the basis of the nature of the sample.[76]
(ii) The procedure to be adopted in cases of dispute.

A1.1 Statistics

Statistical methods aid in the interpretation of data that are subject to random and systematic variability. A set of measurements varying in a haphazard and unpredictable manner about some central value would be exhibiting random variability.[77]

Various books have been published on statistics some of which also cover sampling statistics.[78,79]

A1.2 Normal Distribution

Statistical sampling is based upon the principle that all the particles or portions of the material (population) should have an equal probability of being present in the sample taken. If it were possible to take an infinite number of samples from the lot, and to analyse for a parameter in each of the samples taken, then the frequency distribution of this parameter should follow the Normal Distribution which is illustrated in Figure 21. The graph is symmetrical and falls rapidly on either side.

The normal curve is one of the most useful and widely used distributions because many situations yield data which are approximately normal for practical purposes. Figure 21 illustrates a normal distribution curve for a series of samples of n increments from a population with mean μ and standard deviation σ. The Figure shows that approximately 68% of the sample results obtained are within $\pm$ one standard deviation from the mean. About 95% are within $\pm$ two standard deviations from the mean and 99.7% within $\pm$ three standard deviations. It is clear that normally distributed data will seldom give a value of greater than $\mu \pm$ four standard deviations (approx 1 in 17,000).

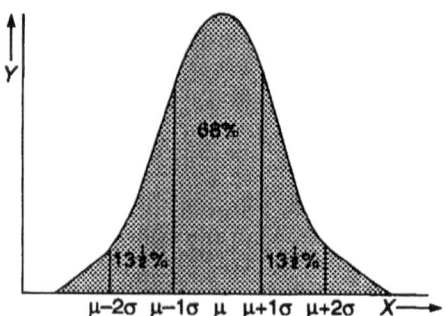

Figure 21

A1.3 Sampling Error

There is generally inherent variation in the composition of the sample eventually analysed. This variation can change depending on the number of sub-sampling stages. This value should be considered and the cumulative variance calculated in order to devise an appropriate sampling plan. In addition there may be an indeterminate error in the analytical method adopted which should be taken into account.[80–83]

Appendix 2: Definitions – List of General Terms Used in Sampling

Aggregate/composite sample: The quantity of product formed by combining and mixing the increments taken from several primary samples (or sub-lots of a consignment).

Batch: A stated quantity presumed to be of uniform characteristics, taken from the consignment, and allowing the quality to be assessed.

Bulk sample: A collected set of a material which does not maintain its individual identity and which may consist of a single pile (coal, fertiliser), a plot of ground (soil), a shipload of grain, or a portion of atmosphere or ocean, all characterised by lack of permanent identifiable units.

Blended bulk sample: A collected set of samples blended together to form a supposedly uniform sample.

Characteristic: A definite property of a material under test which helps to differentiate a particular lot into acceptable and unacceptable items. The difference may be either quantitative (by variables) or qualitative (by attributes).

Consignment: The quantity of product dispatched or received at one time (may consist of a number of discrete packaged units, *e.g.* drums, barrels, cartons, sacks) covered by a particular contract or shipping document. It may be composed of one or more lot or batch.

Composite sample: A blended bulk sample consisting of individual samples taken in proportion to the quantity of material they represent.

Final sample: A representative part of the reduced sample or aggregate sample if no reduction is required. The final sample is the portion sent for possible sub-division for testing, reference, and storage.

Grab sample/spot sample: A single increment taken at a specified location within the bulk (static) or after a specified quantity has passed through the point of sampling or retained as evidence of the material received.

Increment: A small quantity of product taken by a sampling device from each of the sampling units to obtain the primary sample.

Laboratory sample: Primary material delivered to the laboratory.

Lot: The total quantity of material presumed to be of uniform characteristics, taken from a consignment. The material may be one item or a number of items which belong together and which have been produced under uniform conditions. A consignment may consist of one lot or sub-lots which are sampled separately but later combined for assay purposes. The consignment is sub-divided to avoid handling a large primary sample and also if the whole consignment is delivered for sampling on different days in order to reduce sampling errors.

Primary/gross sample: A quantity of material consisting of two or more increments taken directly from a consignment or individual unit of the lot taken as part of the sample taken. When collected together, the individual items retain their identity.

Reduced/divided: Splitting of the aggregate sample into one or more identical portions. The particle size may need to be reduced in the course of reducing the quantity.

Sampling plan: The rules stating the location, the number, and the size of the increments to be taken from a lot. It may state the basis for the acceptance or rejection of the lot.

Sampling unit: A defined quantity of material, for example a packaged unit (container) or at a particular time interval for material in motion, in the consignment.

Test or observation: An operation made in order to measure some property.

Test portion: A quantity of material removed from the test sample and used in a single test or observation. The test portion may be taken from the primary sample or directly from the laboratory sample if no sample preparation is required.

Test sample: The sample prepared from the laboratory sample by further sub-division, mixing, grinding, or a combination of these operations.

References (Bibliography)

1. British Standard BS 5551, Section 2.10, 1993 (International Standards Organisation, ISO 5308, 1992). 'Solid Fertilizers – Methods of Checking the Performance of Mechanical Devices for Sampling of Fertilizers Moving in Bulk'.
2. International Standards Organisation, ISO 6644, 'Cereals and Milled Cereal Products – Automatic Sampling by Mechanical Means', 1981.
3. International Standards Organisation, ISO 5500, 'Oil Seed Residues – Sampling', 1986.
4. Youden, W. J., Statistics in Regulatory Work, *J. Assoc. Off. Anal. Chem.*, 1967, **50**, 1007.
5. MAFF, 'Food Chemical Surveillance', Paper No. 32, HMSO, London, 1993.
6. International Standards Organisation, ISO 3963, 'Sampling from a Conveyor by Stopping the Belt', 1992.
7. Statutory Instruments, Fertilisers (Sampling and Analysis) Regulations 1991, No. 973, Agriculture, HMSO, London.
8. Rainwater, F.H., and Thatcher, L.L., 'Methods for Collection and Analysis of Water Samples', United States Government Printing Office, Washington, 1960.

9. United States Environmental Protection Agency, 'Handbook for Monitoring Industrial Waste Waters', United States Government Printing Office, Washington, 1973.
10. Institute of Civil Engineering,'Safety in Sewers and at Sewage Works', 2nd Edition.
11. Health and Safety Executive Methods, 'Monitoring Strategies for Toxic Substances', EH 42, HSE, HMSO, London, 1989.
12. Health and Safety Executive Methods, 'General Methods for Sampling of Airborne Gases and Vapours', MDHS 70, HSE, HMSO, London 1990.
13. British Standard, BS 5309, Part 1. 'Methods for the Sampling of Chemical Products. Introduction and General Principles; Part 2. Sampling of Gases; Part 3. Sampling of Liquids; Part 4. Sampling of Solids', 1976.
14. Woodgate, B., and Cooper, D., 'Analytical Chemistry by Open Learning', ed. Chapman, W., on behalf of ACOL, John Wiley and Sons, Chichester, New York, Brisbane, Toronto, Singapore, 1987.
15. Smith, R., and James, G.V., 'The Sampling of Bulk Materials', Analytical Science Monograph No.8, The Royal Society of Chemistry, London, 1981.
16. Japanese Industrial Standard, M8100, 'Common Rules for Sampling of Bulk Materials of Mining Products'.
17. Nowak, R., 'International Labmate', 1993, Retsch GmbH, Rheimische Str. 36, 42781 Hann, Germany.
18. Behre, H. A., and Hassailis, M.D., in Taggart, A.F. 'Handbook of Mineral Dressing', Chapman and Hall, London, 1945.
19. Block, E., *Gluckauf*, 1965, **101**, 255.
20. Sporbeck, H., *Z. Anal. Chem.*, 1965, **209**, 60.
21. Jordison, F., *Chem. Eng. (N.Y.)*, 1978, **85**(24), 103.
22. Sellwood, R. M., Proc. 9[th]Commonwealth Min. Metal. Congr., 1969, **3**, 537.
23. British Standard BS 5304, 'Code of Practice – Safeguarding of Machinery', 1988.
24. Perry, J. H., 'Chemical Engineers Handbook', 2nd Edn, McGraw Hill, New York, 1942.
25. Collins, V. G., Jones, J.G., Hendrie, M.S., Shewan, J.M., Wynn, Williams, D.D., and Rhodes, M.E., in 'Sampling – Microbiological Monitoring of Environments', ed. Board, R.G., Lovelock, D.W., Academic Press, London, 1973, pp. 77–110.
26. Anderson, P. W., Murphy, J. J., and Faust, S. D., Proceedings of a Seminar on Design of Environmental Information Systems, ed. Deininger, R.A., Ann Arbor Science Publishers, 1974, pp. 261–281.
27. 'General Principles of Sampling and Accuracy of Results, Methods for Examination of Water and Associated Materials', HMSO, London, 1980.
28. Hutchinson, G. E., 'A Treatise on Limnology', University of Toronto Press, Toronto, 1953.
29. Hutchinson, G.E., 'A Treatise on Limnology', Vol. 1, Wiley, London, 1957.
30. Department of Environment, 'Analysis of Raw, Potable, and Waste Waters', HMSO, London, 1972.
31. Little, A.H., *Pollution Control (London)*, 1973, **72**, 606.

32. Wood, L.B., and Stanbridge, H.H., *Water Pollution Control (London)*, 1968, **67**, 495.
33. Brezonik, P.L., and Lee, G.F., *Air Water Pollution*, 1966, **10**, 549.
34. Howe, L.H., and Holly, C.W., *Environ. Sci. Technol.*, 1969, **3**, 478.
35. Benedek, A., and Najak, A., *Water Pollution Control*, 1975, **31**, (September), 20–24.
36. Devorkin, H., Chass, R.L., and Fudrich, A.P., 'Air Pollution Source Testing Manual', Air Pollution District, Los Angeles, California, 1992.
37. MDHS 25, Health and Safety Executive Methods, 'Isocyanates in Air', HMSO, London, 1987.
38. Health and Safety Executive Methods, 'Arsine in Air', MDHS 34, HMSO, London, 1983.
39. Health and Safety Executive Methods, 'Formaldehyde in Air', MDHS 19, HMSO, London, 1983.
40. Smith, J. R., *J. Air Pollut. Control Assoc.*, 1979, **29**, 969.
41. 'Manual of Analytical Methods', 2nd Edn., Vols. 1–4, ed. Taylor, D.G., US Dept. of Health Education and Welfare, Public Health Service, NIOSH, Cincinnati, Ohio, DHEW Publ. Nos. 75–121, 1975.
42. Bracewell, J. M., and Hodgson, A. E. M., *Int. J. Air Water Pollution*, 1965, **9**, 431.
43. BSI Method for the Measurement of Air Pollution, BS 1747, Part 3, British Standards Institution, London, 1969.
44. Van den Berge, L. P., Devereese, A., and Vanhoorne, M., *Am. Ind. Hyg. Assoc. J.*, 1985, **46**, 693.
45. Gabay, J., Davidson, M., and Donagi, A. E., *Analyst (London)*, 1976, **101**, 128.
46. Morita, H., Mitshuhashi, T., Sakurai, H., and Shimomura, S., *Anal. Chim. Acta*, 1983, **153**, 351.
47. Schenkel, A., and Broder, B., *Atmos. Environ.*, 1982, **16**, 2187.
48. Okita, T., Morimoto, S., and Izawa, M., *Atmos. Environ.*, 1976, **10**, 1085.
49. Grosjean, D., *Anal. Lett.*, 1982, **15**, 785.
50. Thorogood, G., Rapsomamanikis, S., and Harrison, R. M., 'Evaluation of a Method for Measurement of Particulate Chlorine and Gaseous Hydrogen Chloride in the Atmosphere', Contract Report, University of Essex, August 1986.
51. Lewin, E., and Zachau-Christiansen, B., Technical Note, Efficiency of 0.5N Potassium Hydroxide Impregnated Filters for SO_2 Collection, *Atmos. Environ.*, 1977, **11**, 861.
52. 'Methods for the Detection of Toxic Substances in Air, Booklet No.1: Hydrogen Sulphide', HSE, HMSO, London, 1970.
53 BOHS Technology Committee Report, 'Chemical Indicator Tubes for Measurement of the Concentration of Toxic Gases in Air', *Ann. Occup. Hyg.*, 1973, **16**, 51.
54 Leichintz, K., 'Air Analysis by Means of Long-term Detector Tubes', *Drager Review*, Dec. 1977, 40.
55. Slanina, J., Schoonebeck, C. A. M., Klockow, D., and Niessner, R., *Anal. Chem.*, 1985, **57**, 1955.

56. Ferm, M., *Atmos. Environ.*, 1979, **13** 1385.
57. Lamb, S. I., Petrowki, C., Kaplan, I. R., and Simonet, B. R. T., *J. Air Pollut. Control. Assoc.*, 1980, **30**, 1098.
58. Hill, R. H., McCammon, C. S., Saalwaechter, A. T., Teass, A. W., and Woodfin, W. J., *Anal. Chem.*, 1976, **48**, 1395.
59. Health and Safety Executive Methods, Mixed Hydrocarbons C_3 to C_{10} in Air, MDHS 60, HMSO, London, 1982.
60. Fraust, C. L., *Am. Indus. Hyg. Assoc. J.*, 1975, **36**, 278.
61. ASTM, D3686-78, Annual Book of ASTM Standards, Part 26, 1978.
62. Health and Safety Executive Methods, Methods for the Determination of Hazardous Substances, Occupational Medicine and Hygiene Laboratory, London, UK, in series 1981–94.
63. National Institute for Safety and Health, 'Manual of Analytical Methods', 2nd Edn, DHEW, published 1990 No. 75–121 (1975); 3rd Edn., DHEW No 84–100 (1984 revised 1987).
64. Guenier, J. P., Simon, P., Delcourt, J., Didiorgean, M. F., Lefevre, C., and Muller, J., *Chromatographia*, 1984, **18**, 137.
65. Health and Safety Executive Methods, 'Glycol Ethers and Glycol Acetate Vapours in Air', MDHS 23, HMSO, London, 1988.
66. Health and Safety Executive Methods, 'Benzene in Air', MDHS 22, HMSO, London, 1990 (revised).
67. Billings, W. N., and Bidelman, T. F., *Environ. Sci. Technol.*, 1980, **14**, 679.
68. Krist, K. J., Pellizzani, E. D., Walburn, S. G., and Hubbard, S., *Anal. Chem.*, 1982, **54**, 810.
69. Pellizzari, E.D., Gutknecht, W.F., Cooper., S., and Hardison D., 'Evaluation of Sampling Methods for Gaseous atmospheric Samples', EPA-600/3-84-062 Environmental Protection Agency, USA, 1984.
70. Health and Safety Executive Methods, 'n-Hexane in Air', MDHS 74, HMSO, London, 1992.
71. Health and Safety Executive Methods, 'Acrylonitrile in Air', MDHS 55, HMSO, London, 1986.
72. Health and Safety Executive Methods, 'Formaldehyde in Air'. MDHS 78, HMSO, London, May 1994.
73. Health and Safety Executive Methods, 'Protocol for Assessing the Performance of Diffusive Sampler', MDHS 27, HMSO, London, 1987.
74 'Detection and Measurement of Hazardous Gases', ed. Cullis, C.F., and Firth, J.G., Heinemann, London, Exeter, 1981.
75. Strickland, J. D. H., and Parson, T. R., 'A Practical Handbook of Seawater Analysis', Fisheries Board of Canada, Ottawa, 1968.
76. 'Sampling Inspection Statistical Research Group, Columbia University', ed. Freeman, H., Friedman, M., Mosteller, F., and Wallis, W., 1st Edn, 1948.
77. Kendale, M. C., Stuart, A., and Griffin, 'The Advanced Theory of Statistics', 1958.
78. Moroney, M. J., 'Facts and Figures', 4th Edn, Penguin, Harmondworth, 1969.
79. Davies, O.L., and Goldsmith, P.L., 'Statistical Methods in Research and Production', 4th Edn, Oliver and Boyd, Edinburgh, 1972.

80. Stuart, A., 'Basic Idea of Scientific Sampling', Statistical Monographs and Courses No. 4, ed. Stuart, A., 2nd Edn, Griffin, 1976.
81. 'Quantifying Uncertainty in Analytical Measurement', Eurachem., Version 6, January 1995.
82. 'Guide to Statistical Interpretation of Data', BS 2846, Part 5, ISO 3301, 1977.
83. Pearson, K., 'Biometrica Tables for Statisticians', 4th Edn, Arnold, 1976.